Poison

Originally published in 2019 by André Deutsch, an imprint of the Carlton Publishing Group
This edition published in 2020 by Welbeck, an imprint of Welbeck Publishing Group
20 Mortimer Street
London W1T 3JW

Text © Welbeck Non-fiction Limited 2020
Foreword © Sophie Hannah 2020
Design © Welbeck Non-fiction Limited 2020

All rights reserved. This book is sold subject to the condition that it may not be reproduced, stored in a retrieval system or transmitted in any form or by any means, electronic, mechanical, photocopying, recording or otherwise without the publisher's prior consent.

A CIP catalogue for this book is available from the British Library.

ISBN 978 0 233 00611 6

Printed in Dubai

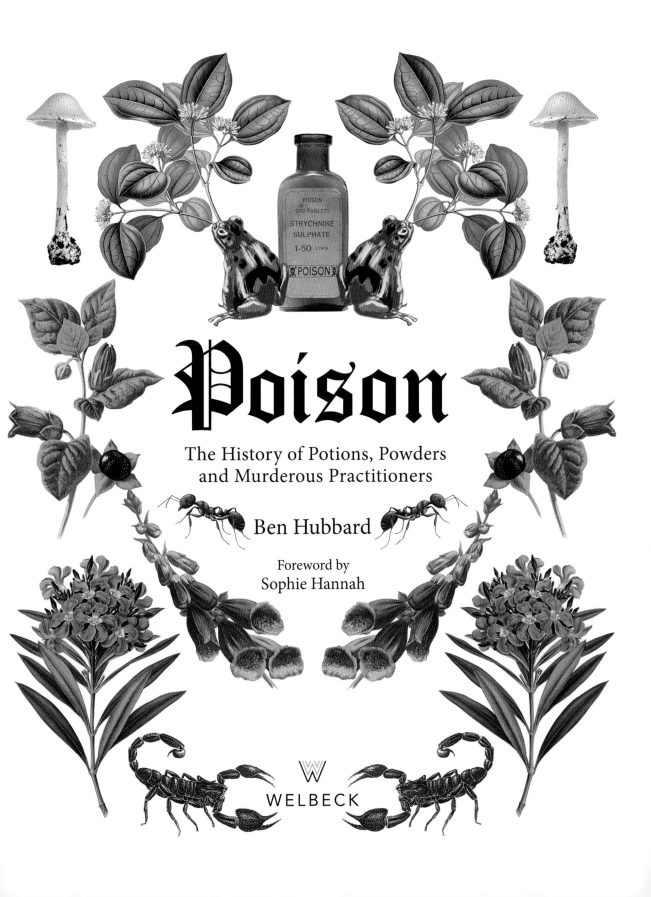

Poison

The History of Potions, Powders and Murderous Practitioners

Ben Hubbard

Foreword by
Sophie Hannah

Contents

Foreword 6

Introduction: Poison – a Recurring Story 8
Poisonous Plants and Predators, Elements of the Earth, The Science of Poisoning

1
Poisons of the Ancient World 16

Sentencing Socrates, Hemlock, Killing Cleopatra, Mithridates' Mass Suicide, Belladonna, Agrippina and Nero: A Family Affair, Aconite, Lead, Purging Qin Shi Huang, Mercury, Temptress Empress Wu

2
Medieval and Renaissance Poisons 46
The Bloated Borgias, Arsenic, The Tofana Trap, The Catherine de' Medici Influence, Executing Elizabeth I, Opium

3
Seventeenth- and Eighteenth-Century Poisons 66
Dosing Marquis de Sade, Cantharides, Affair of the Poisons, Salem Witch Trials, Ergot, The Jamestown Poisonings, Datura, Murdering Mary Blandy

4
Nineteenth-Century Poisons 88

Marrying Florence Maybrick, Bradford Sweets Poisoning, Mothering Mary Ann Cotton, The Matchstick Girls, Phosphorus, Villain William Palmer, Strychnine, Thomas Cream's Chloroform, Chloroform

5
Twentieth-Century Poisons 112

Assassinating Grigori Rasputin, Cyanide, Hitler's Cyanide Command, Alan Turing Tragedy, The Jonestown Massacre, Hawley Crippen's Homeopathy, Georgi Markov's Umbrella Death, Ricin, The Tokyo Sarin Poisonings, Sarin, Harold Shipman's Sentence, Threatening George Trepal, Teenage Japanese Poisoner, Thallium

6
Twenty-first Century Poisons 148

The Washington Anthrax Attacks, Anthrax, Iceman Richard Kuklinski, Alexander Litvinenko Assassination, Polonium-210, Viktor Yushchenko Disfigurement, Dioxin, Salisbury's Sergei and Yulia Skripal, Novichok

Index 172
Credits 176

Foreword

"For it is a devilish thing to do – to poison a man in cold blood. If there had been a revolver about and she'd caught it up and shot him – well, that might have been understandable. But this cold, deliberate, vindictive poisoning – and so calm and collected." That's a quote from Agatha Christie's *Five Little Pigs*. In fact, it later turns out that the decision to poison Amyas Crale was anything but cold and calm. It was rather a hot, passionate poisoning, as poisonings go ... but I will say no more, as I do not wish to spoil one of the best detective novels of all time for anyone who has not yet read it.

Agatha Christie was a poison expert, having worked as a hospital pharmacist as a young woman, and it seems that becoming her continuation writer (or sidekick, as I prefer to think of it!) unleashed my own fictional poisoning capacity. In my own crime writing, I had never used poison as a murder method until I wrote my first Poirot mystery, *The Monogram Murders*, in 2013. Suddenly, I seemed to need it: cyanide, to be specific. I used poison again in my second Poirot novel, *Closed Casket*: strychnine this time. Was it the historical 1930s setting that liberated my inner poisoner? I don't think it can have been only that. Soon after writing *Closed Casket* I published a contemporary crime story, *Bully the Blue Bear*, in which a food substance to which the victim is allergic is used as a personally-targeted poison. The murderer in that story would have baulked at using cyanide or strychnine even if she could have obtained them, because those are well-known Poisons with a capital P, but somehow carrying a few pieces of boiled egg into a building and dropping them into a particular lunchbox did not feel quite so reprehensible to this character – she did not feel quite so much like a murderer – even though she knew that she was actively seeking to cause a death.

This, I think, fits well with the psychology of the poisoner. It's a murder method that allows for dishonesty and hypocrisy in abundance, and it is surely the people-pleaser's weapon of choice – your victim never need know you're killing him. You can drop the poison into the relevant receptacle and then arrange to be elsewhere when the unpleasant choking and dying happens. If you are squeamish, you can avoid witnessing a death, and you can also dodge the disapproval and hatred of your victim. Have you ever met someone (such people are numerous) who needs to be liked even by people they loathe? If you want to kill someone without them suspecting that you're anything but their most loyal friend and supporter, that's completely feasible if you use poison. You can be miles from the scene of the crime by the time your target dies, whereas if you shoot or stab your victim, the necessary proximity involved means there is a strong chance of being seen and recognized for the killer you are.

But there are disadvantages for the people-pleasing poisoner too – because people-pleasers are, essentially, manipulators who seek to control the opinions and behaviours of others. If you're a manipulative control freak, you won't want simply to drop some poison into a glass and wander off, assuming your victim will die – there would be too many uncertainties: what if they spill that glass? Or pass the drink containing the poison to someone else, as happens in Agatha Christie's *The Mirror Crack'd From Side to Side*?

There's another type of murderer (or crime writer) who might avoid poison as a method: people who don't like illness. In cases of murder by gun or knife, your victim is perfectly healthy one moment, and then suddenly they're dead. There's no body that grows gradually sicker. Death by poison can look like a speeded up version of what might happen in a case of terminal illness, and to some tastes it might seem preferable to go straight from healthy to dead – if you abhor illness and want your victim out of the way but not for them to suffer unnecessarily, for example. Those of us who write murder mysteries more for the mysteries than for the murders might not want to waste unnecessary time on choking and vomiting and horrible things like that when there's a baffling psychological puzzle on which we would prefer to focus our attention.

I'll admit it. I'm talking about me. Here is a slightly embarrassing story: I was once asked to read and offer a praise-ful cover quote for a crime novel about a woman who was on trial for murdering her terminally ill husband. She admits to killing him, and her defence is that it was a mercy killing, to end his suffering. I said to the publisher, "I have to ask: is there a twist? Does it turn out that he's not terminally ill, and she just hated him and wanted him dead?" "No," said the slightly bemused publisher. "He really was dying, and she loved him, and it really is a mercy killing." "In that case I can't read the book," I said. 'I'd find that story too upsetting.'

My third Poirot novel, *The Mystery of Three Quarters*, did not involve poison. I'm currently planning my fourth, and I haven't yet decided how the victim will die. Like Poirot, I'm an order and tidiness freak and there's something messy about poison. I can't help thinking that murder should be quick, honest and face-to-face, and of course, never committed in the first place, apart from in the pages of novels. If you're a crime reader, true crime fanatic or aspiring crime writer, there's no doubt that you'll find in this book a wide range of fascinating poison options. Just make sure you wipe your fingerprints off the cover after reading.

Sophie Hannah

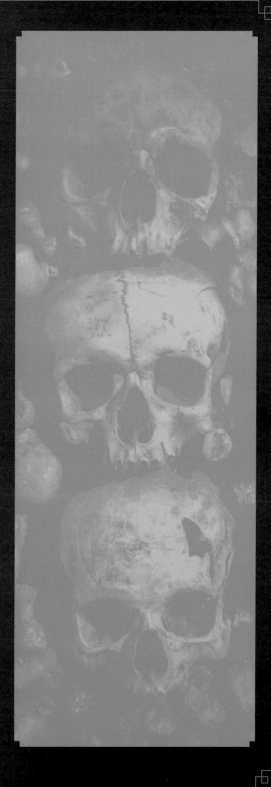

Introduction: Poison – a Recurring Story

In 2018, the world was stunned by reports of a geo-political poisoning that seemed straight from the pages of a cold-war thriller. In the small town of Salisbury, England, former Soviet spy Sergei Skripal and his daughter Yulia were poisoned by a military-grade nerve agent called novichok.

The poison, which was sprayed onto the Skripals' front door handle, causes respiratory collapse, as the victims' lungs fill with their own body fluid. The Skripals spent weeks in a coma and were lucky to survive.

News stories about murder through poisoning have a macabre allure. Poisoning, after all, is a dark art as old as human history itself. However, novichok – which was developed in secret laboratories deep in the Soviet Union – was only the latest in humanity's search for the perfect poison. In many ways novichok came close: it is colourless, odourless, simple to disguise and easy to administer. But there was nothing subtle about the message behind the Skripal poisonings: it was a warning to all Russian political dissidents that a violent death could strike at any time.

Earlier searchers for the perfect poison usually required something less obvious. The best poisons could be hidden in a ring or a make-up compact and cause symptoms that mirrored common illnesses or diseases. Forensic investigations for poisons did not begin until the early nineteenth century, the era known as the Golden Age of poisoning.

During the Golden Age, arsenic was so prevalent that you could buy it as easily as a loaf of bread. There was arsenic in wallpaper, children's toys and soap, so it was little wonder that people often became sick or even dropped dead from accidental arsenic poisoning. This was good cover for poisoners, who often slipped arsenic into a loved one's tea; the introduction of life insurance provided some extra incentive. However, arsenic taken at lower doses had long been considered a tonic that could boost the immune system and prolong life.

The ancient king Mithridates was a great advocate of this theory. He ingested a daily concoction of poisons to protect himself against possible assassination. If the poisons were taken at low doses, Mithridates reasoned, his body would build up a resistance to them and he would become immune. It proved an appealing theory with an enduring legacy: a version of the king's *Mithridatium* was still being sold up to the late eighteenth century.

It is no coincidence that the history of poison has run in tandem with the history of medicine: the two are often born of the same substance. It was the Swiss physician Paracelsus who showed that the difference between a poison and a medicine is often simply the dose. "All things are poison and nothing is without poison; only the dose makes a thing not a poison," Paracelsus wrote in the 1530s, and in doing so founded the modern scientific field of toxicology. His principle states that all chemicals – even water and oxygen – can be poisonous in high doses.

The theory that a medicine at a low dose is also a poison at a higher one has been understood from the time of our earliest civilizations. The ancient Egyptian *Ebers Papyrus* is one of the world's oldest medical documents and contains recipes for over 700

ABOVE: Rodine was a widely-available British rat poison made of bran, molasses and phosphorus. Phosphorus poisonings were said to decrease after it was banned in 1963.

formulas and remedies. Some of these are magical and spiritual in nature; others suggest the ingestion of honey, sycamore fruit and dates for ailments such as urinary problems. The 1534 BC papyrus also contains recipes containing poisons, including hemlock, aconite, opium and antimony.

Egypt's last pharaoh, Cleopatra, reportedly experimented with poisons on condemned criminals and recorded the results. Her own end was famously said to have come from the venom of an asp bite. Cleopatra's death, in defiance of capture by Rome's Octavian, reminds us that poison is one of history's great levellers. It is a murder weapon that does not require brawn to administer: throughout this book both women and men have murdered people using poison.

Historically, women have also been great manufacturers of poison substances. In the seventeenth century, alleged sorceress "La Voisin" sold many poisons to nobles at the court of Louis XIV; Italy's Giulia Tofana invented a poison so infamous it bore her name and was perhaps responsible for the deaths of over 600 people. Often it was used by wives to free themselves from abusive husbands.

Poisoning is a continuous dark thread in the fabric of human history. It reflects our advances in science, technology and social thinking; it also holds a visceral fascination. Murder through poisoning is personal and unambiguous; there can be no confusion about its intent. The symptoms are shocking, curious and myriad: victims can die during extreme convulsions, or have their bodies devoured from the inside out. Others suffocate as their respiratory systems collapse and blood streams from their noses, mouths and eyes. Cyanide poisoning can contort a dead victim's face into a horrible, sardonic grin.

The dreadful history of poisoning goes back to our earliest beginnings and still plays out on the world stage today. It is about our best and worst impulses: the desire for knowledge and improvement, but also the need to destroy others. It is a profoundly human story.

Poisonous Plants and Predators

For many centuries, most poisons available to murderers came from plants and animals. The natural world is a toxic battleground where poisons are used in both defence and attack.

Plants use powerful toxins to protect themselves against hungry predators. Venomous animals use lethal toxins to kill their prey. Nature, therefore, offers a smorgasbord of toxic substances for humans with malevolent intent.

PLANTS

The first people turned to plants for their poisons. Human understanding of plants – both for good and ill – stretches back to our earliest beginnings. Our near relatives, such as apes and chimpanzees, regularly ingest pharmacologically-active plants to alleviate stomach aches and other ailments. As humans evolved from primitive hominids to uprights, we too must have inherited some of this atavistic knowledge. However, a plant offering medicinal qualities can also be deadly at higher doses. The first humans trying to treat themselves with plants must have occasionally dropped dead. Accidental poisoning has always been a risk with plants.

Determining the correct toxicity of a plant remained a notoriously tricky process until science became capable of artificial synthesis. This is because the potency of a plant's poisons, called phytotoxins, can vary considerably; even two identical plants grown in the same location can have toxins of different strengths. These variables, combined with a victim's personal biological response to a poison, have added to the "trial and error" process of early poisoning stories. As we will see, attempted poisonings using the plants belladonna, datura and hemlock have had unusual, unexpected or at times underwhelming results.

*ABOVE: Henbane (*Hyoscyamus niger*), aconite (*Aconitum *species) and belladonna (*Atropa belladonna*) are plants that can cause hallucinations and the sensation of flying. In medieval times, female healers applied an intoxicant ointment extracted from these plants to their genitals via broomstick handles. Witch mythology is thought to stem from this.*

FUNGI

The poisons produced by fungi are called mycotoxins and are among the most potent in the world. Fungi range from minuscule, single-celled moulds to mushrooms and toadstools. Toadstool poisoning is extremely common because people often mistake toadstools for edible mushrooms. Poisoning happens frequently with the death cap toadstool (*Amanita phalloides*) and the fly agaric mushroom (*Amanita muscaria*).

The fly agaric has a long history not only as a poison but also as an hallucinogenic intoxicant. The indigenous people of Siberia used fly agaric in ceremonial rituals. A shaman would swallow the mushrooms and others would drink his urine to experience its psychoactive effects. This process would filter out some of the negative toxic effects of fly agaric, which include sweating, nausea and itching. Fly agaric is known for the unpredictability of its effects: symptoms associated with an overdose include confusion, hallucinations and spasms; but these may also be experienced in milder form during a fly agaric 'trip'. Another, smaller fungal mould called ergot has similar effects to fly agaric. However, as ergot is a harmful fungus that grows on rye and barley, it is never ingested intentionally. The results of ergot poisoning can have implications more serious than just the medical, as was seen during the Salem witch trials (pages 78–79).

ANIMALS

There are thousands of poisonous creatures throughout the animal kingdom, with birds being the only exempt group. Animal poisons are called zootoxins and they exist in the bodies of certain creatures as a method of defence. The cane toad (*Rhinella marina*), for example, has a highly toxic hide and if threatened can excrete a poison known as bufotoxin. The blister beetle (family Meloidae) contains a poison known as cantharidin, which it uses to protect its eggs. Cantharidin is a poison sometimes incorrectly taken as an aphrodisiac, often with toxic results. A famous case involved the Marquis de Sade (pages 68–71).

Other animals use poison to attack. Snakes, scorpions and spiders are the most notorious of these venomous creatures, with snake bites alone accounting for at least 100,000 fatalities annually worldwide. Vipers, cobras and taipans are among the world's most poisonous snakes. Their venom – injected via hypodermic-like fangs – can break down the tissue at the site of the bite, cause paralysis, and result in respiratory and cardiac failure. Many believe that Cleopatra of Egypt committed suicide via a cobra bite to the breast. This theory is examined on pages 22–25.

ABOVE: The fungi fly agaric has been linked with the origin of the Santa Claus story. It is an intoxicant that causes the sensation of flying; it is consumed by reindeer; and it bears the traditional red and white colours of Santa Claus.

Elements of the Earth

As human history advanced, new poisons were discovered that were not from plants or animals. These initially came from earth elements, and later, bacteria and man-made substances previously unseen or unknown by humans.

These toxins are the most lethal substances that we know, and have been created with mass murder in mind.

ELEMENTS

There are around 80 elements in the Earth's crust. At low levels, these elements are harmless; many of them can be found in trace amounts in the human body. In higher doses, however, the elements provide some of the most famous poisons known to history: arsenic, antimony, mercury, lead and thallium. Traditionally, these elements were used in dyes, cosmetics and rat poisons, but also medicines. The ancient Egyptians recommended antimony for skin conditions, while mercury was used in pharmaceutical pills, such as laxatives, well into the 1970s. Mercury was also taken as a daily tonic by China's first emperor, Qin Shi Huang (pages 40–41).

The use of lead in ancient tableware caused the chronic poisoning of many people of the Roman Empire, which some today believe contributed to its demise (pages 38–39). Arsenic is perhaps the most famous poison of all, and its use by murderers appears throughout this book. It is interesting to note arsenic's place as a medicine: in the mid-nineteenth century the "arsenic eaters" of Styria, Austria, were found to consume arsenic in small quantities and swore by its health benefits. This gives further credence to physician Paracelsus's theory that all substances are both medicines and poisons, depending on the dose.

ABOVE: Nineteenth-century dvertisement for Dr Mackenzie's Arsenical Soap, which claims to cure spots, pimples and produce a lovely complexion. The ingredients include zinc and arsenic.

BACTERIA

Humans are constantly at war with harmful bacteria. Too small to be seen with the naked eye, they are transported into our bodies through food, water and air. Once inside, the bacteria grow and multiply at an alarming rate. As they do this, they produce poisonous proteins known as exotoxins. These toxic chemicals are released constantly, even during a bacterium's death throes as it is destroyed by the body's immune system. Bacterial poison attacks humans at a cellular level, by either disrupting the normal activity of cells or simply destroying them completely. This is sometimes done by cracking open the cell's outer membrane wall and spilling its contents before stopping its protein synthesis and blocking messages at its nerve junctions. Weaponized bacteria, such as anthrax, make a deadly poison that can devour the body from within. More can be read about anthrax poisoning on pages 152–53.

MAN-MADE

Manufactured chemicals have a variety of important uses in the modern world, such as in agriculture and pest control. However, the same chemicals developed to kill insects, rodents, fungi and weeds have often been turned on humans as well. To do this, chemical manufacturing plants were simply turned into laboratories for producing weapons of mass destruction. The rat poison cyanide is perhaps history's most chilling example. Zyklon B was a cyanide-based pesticide used by Nazi Germany to murder over one million people in extermination camp gas chambers (pages 116–17).

In the twenty-first century, mass murder on Tokyo's subway system was attempted with the poisonous gas sarin. Sarin is an example of a modern poison developed specifically as a weapon to kill humans; the nerve agent novichok is another. Novichok was one of the weapons developed in secret laboratories in the Soviet Union during the Cold War and not unleashed until 2018 (pages 166–71). Attempted assassinations by novichok and the radioactive element polonium represent poisons as the murder weapons of today's world.

ABOVE: The Black Death that ravaged Europe in the fourteenth century was caused by a bacterial outbreak known as the bubonic plague. Between 25 and 40 million people died as a result.

ABOVE: Used prominently in the extermination camp Auschwitz-Birkenau, Zyklon B was administered as pellets through vents in the gas chamber to murder those inside.

The Science of Poisoning

In its simplest terms, a poison is a toxic substance that causes injury, illness or death when absorbed by an organism. There are numerous poisonous substances, which come from plants, animals or elements in the Earth's crust.

Some other poisons are man-made, either discovered by chance or developed in top-secret weapons laboratories. Poisons come in various physical forms, including solids, liquids, gases, vapours and aerosols. A poison's physical form, its dosage and method of delivery are the factors that determine its exposure and effects on the human body.

ROUTES OF EXPOSURE

The route a poison takes into a person influences how quickly it will act, which parts of the body it will attack, and the victim's chances of survival. A substance can enter the body in four main ways: by ingestion, inhalation, absorption or injection.

Ingestion: For poisons in solid form, ingestion is the main route of entry. Because this is the same route as digesting food, the body has some natural defences to expel the poison. Insoluble poisons are simply pushed along the gastrointestinal tract and excreted in the normal way. Soluble poisons, on the other hand, can penetrate the gastrointestinal lining and travel into the bloodstream. They are then transported to the liver and other organs, where they can cause the most damage.

Inhalation: For poisons in the form of vapours, gases or aerosols, inhalation is the main route of entry. Once inhaled, these poisons travel into the respiratory tract and then into the bloodstream through the lungs. This makes the effect fast and dangerous: blood from the lungs goes directly to the heart and is then pumped around the rest of the body. The brain is one of the first organs to receive this blood, but often the chemicals inhaled in poisons end up travelling to the organ for which they seem to have most affinity: this is known as the "target organ".

ABOVE: The two main families of front-fanged venomous snakes are elapids and vipers.

Absorption: Liquid poisons that can be absorbed through the skin or eyes have a doubly dangerous effect. This is because the corrosive chemicals in the poisons can cause extreme tissue damage as they pass through the skin and into the bloodstream. Once in the bloodstream, the poisons are transported to the internal organs. Human eyes are particularly sensitive to toxic substances and can be irreparably damaged if a poison is administered in this way.

Injection: Poisons that enter the body through a hypodermic needle or the fangs of an animal are particularly deadly. Aside from inhalation, injection is the fastest way into the bloodstream. The poison simply bypasses many of the body's natural defence mechanisms. Snake bites can also destroy the tissue around the site of the bite and prevent healing agents in the blood from reaching the wound.

THE BODY'S REACTION

Once a poison has entered the bloodstream, it attacks the body at the cellular level. Some poisons do this by disrupting the messages that pass between the nerves and muscles; these poisons stay outside the cells. Most poisons, however, pass through a cell's outer membrane into its centre. This can happen in several ways: some poisons force their way through the outer membrane; others dock with molecules that travel through cellular tunnels in the membrane. The most dangerous poisons are lipid-soluble: these can simply pass through the lipid-constructed membrane anywhere they like. Once inside a cell, a poison can cause maximum damage to the victim's biological systems. It can do this by stopping the cell synthesizing DNA (which carries the cell's genetic code) and RNA (which carries out certain vital functions in a cell) – its main purpose – or blocking the enzymes that enable the cells' different reactions. Some poisons pose as helpful enzymes, so they can destroy the molecules within a cell. Other poisons block the cell's energy supply, so it starves. Some poisons leave the cell's outer membrane open so fluid flows in and fills it up until the cell bursts.

ABOVE: *Over 40 years since the Vietnam War, children are still born with defects caused by Agent Orange. Its main ingredient, dioxin, remains in the country's ecosystem.*

ACUTE OR CHRONIC DOSAGE

Poisoning can occur from either acute or chronic exposure to a toxic substance. Acute poisoning is exposure to a poison on one occasion, often delivered as a single dose. Chronic poisoning is exposure to a toxic substance over a prolonged period of time; this can be anything from a few days to several decades. The toxic symptoms of asbestosis poisoning, for example, can be latent for over 20 years before emerging.

Many poisons are equally effective as both acute and chronic poisons, depending on the dosage. The quantity of the poison, the frequency of the dosage, and the speed at which it becomes active all combine to determine a poison's overall effect. In medical parlance, a dosage of poison that that would kill 50 per cent of a test group is known as the lethal dose, or LD50.

Chapter 1: Poisons of the Ancient World

When was the first human murdered with poison? He or she has no name, but remains a figure lost in our primordial past. Archaeology shows us that poison was first used by our earliest ancestors on spear points and arrows.

This would have brought down a large mammal more efficiently than a prolonged bout of butchery with axes and clubs. When meat was scarce, experimentation showed cooking could eliminate the poisons in plant roots and tubers. The science here was just trial and error; many ancients must have died to provide recipes for our safe eating today.

As hunters and gatherers progressed from being nomads to settlers, a new need for poisons emerged. People living together at close quarters attracted vermin and pestilence, and rat poison was an efficient solution. But how long before this toxic concoction made it into the cooking pot of a rival, a hated family member, or a faithless lover?

As villages became towns and cities, written records appeared of poison as a means of murder. The ancient Sumerians, Akkadians and Egyptians all knew about plant poisons, including opium and belladonna. Their texts suggest vinegar to counteract a poison, a practice still in use today. In the east, Shennong, the mythical founder of Chinese medicine, identified hundreds of poisonous herbs by personally testing their properties; his transparent body allowed him to see the physical results of these toxins.

"Toxin" comes from the ancient Greek *toxikón*, or more particularly *toxikós*, which pertains to poison arrows. In Homer's *Odyssey*, Odysseus famously dips his arrows in the poison hellebore, and the Greeks and Trojans use poison arrows and spears against each other in the *Iliad*, also attributed to Homer. The mention of poison in *Odyssey* (eighth century BC) shows how far back its legacy stretches in Greece.

But these men of the Bronze Age belonged to an era of savagery, according to Roman poet Ovid. His 8 AD poem *Metamorphoses* reports the use of poison at this time for familial murder: "Husbands longed for the death of their wives, wives for the death of their husbands. Murderous stepmothers mixed deadly poisons, and sons inquired into their father's years before their time."

By the time of the Roman Empire, murdering family members had gone mainstream, especially among the ruling class. Then, it was not altogether unexpected for an emperor such as Nero to poison his step-brother and mother; it was a habit, after all, that he had picked up from her.

However, it is not in Rome but in the earlier, allegedly more civilized period of classical Greece that this chapter begins. It centres on classical Athens and the foremost philosopher of the ancient world, Socrates.

OPPOSITE: Homer's Odysseus used poisoned arrows to dispatch his wife Penelope's suitors after returning to Ithaca from the Trojan War.

Sentencing Socrates

One day in 399 BC, the philosopher Socrates stood before a jury of 500 Athenians. He was accused of "refusing to recognize the gods recognized by the state" and "corrupting the youth".

A guilty verdict could mean death. After six hours of arguments from Socrates and his accusers, the jury voted by placing "guilty" or "not guilty" discs into an urn. The verdict: guilty by 280 to 220.

The jury was asked to decide on Socrates' penalty, which his accusers argued should be the death penalty. When asked for his opinion, Socrates impudently suggested that he receive a government salary and free dinners for the rest of his life. His second suggestion was that he pay a small fine.

Athenians had come to expect no less from Socrates, a man who was widely hated by the city-state. According to Plato, his student and biographer, Socrates had become the "gadfly" of Athens when the oracle at Delphi had allegedly said no one was wiser than he. Socrates regularly annoyed his fellow citizens: he enraged the *polis*, the political community, by criticizing democracy, a political system invented in Athens; and he embarrassed people with his Socratic method of questioning, which invariably exposed ignorance. Socrates was also associated with the Thirty Tyrants, who had briefly overthrown the Athenian government.

But now Socrates was about to receive his comeuppance. His sentence was execution, to be administered by his drinking a cup of hemlock. Plato was not present at the execution, but his *Phaedo* recalls it from eyewitness accounts told to him. This is what happened after Socrates had drunk the poison:

"Socrates walked around until he said that his legs were becoming heavy, when he lay on his back, as the attendant instructed. This fellow felt him, and then a moment later examined his feet and legs again. Squeezing a foot hard, he asked him if he felt anything. Socrates said that he did not. He did the same to his calves and, going higher, showed us that he was becoming cold and stiff. Then he felt him a last time and said that when the poison reached the heart he would be gone. As the chill sensation got to his waist, Socrates uncovered his head (he had put something over it) and said his last words: 'Crito, we owe a cock to Asclepius. Do pay it. Don't forget.' … after a while he gave a slight stir, and the attendant uncovered him and examined his eyes. Then Crito saw that he was dead, he closed his mouth and eyelids."

So ends the account of one of the most famous early poisonings in history, and with it Socrates, the father of Western philosophy.

OPPOSITE: *Jacques-Louis David's 1787* The Death of Socrates *juxtaposes the emotion of his followers against the condemned philosopher's stoicism. The 70-year-old Socrates is somewhat fitter and handsomer than statues from antiquity show.*

Hemlock

Poison hemlock (Conium maculatum) is a highly toxic plant that contains coniine, a volatile neurotoxin that is harmful to the human central nervous system. It is also known as the devil's porridge, poison parsley and muskrat weed.

OVERVIEW:

Poison hemlock is a shiny, bright green plant that grows to around 2.5 metres tall, often has purple spots on its stem, and has white flowers in spring. Said to resemble non-poisonous plants such as parsley or carrot, every part of hemlock contains the toxic alkaloid coniine, including its seeds, flowers, leaves and fruit. Even in small amounts coniine can be rapidly fatal, with symptoms appearing within 30 minutes of ingestion. Secondary poisoning can also take place: quail that eat hemlock seeds can pass the poison on to a human who consumes their flesh; diarrhoea, vomiting and paralysis can occur three hours later. Famously prescribed as a suicide and execution poison in ancient Greece, hemlock was used to kill the philosopher Socrates (page 19). However, it has been shown that hemlock can produce violent convulsions in those who have ingested it, which is at odds with Socrates' reportedly gentle death. It has been suggested that the poison given to Socrates was a mixture of hemlock and another substance such as opium, which mitigated the violent effects of the poison. What he was given was reported to create the sensation of a creeping numbness from the extremities inwards, until the poison stopped the philosopher's breathing.

TOXIC EFFECTS:

The coniine alkaloid found in hemlock has the same structural make-up as nicotine: it disrupts the mechanism of the central nervous system by acting on the nicotinic acetylcholine receptors. It can also block impulse transmission to the muscles, which leads to paralysis of the respiratory muscles and lack of oxygen to the heart and brain. Because this alkaloid is present in every part of hemlock, even touching the plant may cause a skin reaction in some people. Hemlock poisoning can cause significant pain and a distressing death, contrary to accounts provided by the ancient Greeks.

SYMPTOMS:

The symptoms of hemlock poisoning often begin with trembling, gastric pains, increased salivation, muscle weakness, rapid heart rate and loss of speech, followed by convulsions, paralysis, central nervous system depression, respiratory failure, asphyxiation, blindness, coma and death.

TREATMENT:

There is no antidote for hemlock poisoning, but pumping the stomach shortly after it is ingested may prevent serious symptoms from appearing. Otherwise, the drug diazepam may be given for convulsions.

FAMOUS POISONINGS:

- In 2006, an English gardener from Devon imitated Socrates' death by ingesting leaves from the hemlock plants growing around his property. His dead body was found a week later. An autopsy revealed that the man's respiratory and nervous systems had become paralysed, which had in turn restricted movement in over 70 per cent of his throat; asphyxiation followed.

OPPOSITE: *A scientific illustration of the poison hemlock plant (Conium maculatum).*

Killing Cleopatra

The end for Cleopatra, Egypt's last pharaoh, began with her failure at the sea battle of Actium in Greece in 31 BC – a victory for Cleopatra's foe Octavian, Rome's first emperor.

Sensing that defeat was near, Cleopatra's flagship famously fled the battlefield. The Roman leader Mark Antony, her lover and partner against Rome, dutifully sailed after her. Their fate as history's most famous suicidal lovers was now sealed.

Cleopatra and Mark Antony had little choice but to flee Actium for her royal capital in Alexandria. Here, they shut themselves behind their palace gates and prepared for the inevitable. The final defeat of Mark Antony was all that stood between Octavian and his full control of Rome. Cleopatra had been a convenient scapegoat for war with Antony; she was painted in Rome as an Eastern whore and temptress who had seduced Antony, Julius Caesar's former right-hand man, and turned him against his former ally Octavian, Caesar's heir.

Antony and Cleopatra had gambled everything on a victory at Actium: now Octavian was closing in for the coup de grâce. Antony rallied what forces remained loyal to him, while also attempting to parley with the approaching Octavian; he received no reply. Cleopatra was also trying the diplomatic route after her remaining ships had been burned in Alexandria harbour; escape was now impossible. Her diplomats brought lavish gifts to Octavian and pleaded with him to let her and Antony live unmolested in Egypt. She also requested that her royal line be allowed to continue under her children, born both of Mark Antony and of Julius Caesar, her former lover.

Messages went to and fro. One of Octavian's envoys told Cleopatra that no deal could be done while Antony was still alive; it was up to her, therefore, to kill him. However, suspecting foul play, Antony had the man flogged and sent back to Octavian without an answer. Cleopatra, perhaps sensing that Antony was a liability, began considering a final end through suicide. This exit was first suggested after Actium, when Antony had nearly taken his life in despair.

Cleopatra herself had a fascination with poisons and a working knowledge of their effects. She had tested various plant and animal toxins on prisoners condemned to death and carefully recorded the results. According to Roman historian Plutarch, Cleopatra discovered that "the bite of the asp alone induced a sleepy torpor and sinking, where there was no spasm or groan, but a gentle perspiration on

ABOVE: The bite of an Egyptian cobra, sometimes known as "Cleopatra's Asp", is venomous enough to kill an elephant in under three hours.

the face, while the perceptive faculties were easily relaxed and dimmed … as is the case with those who are soundly asleep."

Antony's end came after a final battle with Octavian outside Alexandria ended on 1 August 30 BC, when Antony's remaining men surrendered. On hearing the news, Cleopatra locked herself in her mausoleum and sent word to Antony that she had killed herself. At this news, Antony stabbed himself in the stomach with his sword, the traditional manner of suicide used by fallen Roman generals.

Cleopatra, however, was not dead, but instead playing for time and the possibility of a way out for herself and Egypt. But it was a desperate last hour for the pharaoh: Octavian's troops had entered the palace and taken the royal children hostage. When the two leaders came face to face she told Octavian bluntly that she would not be paraded through Rome in a triumph. But Octavian would give her no assurances. Soon afterwards a rumour spread that Octavian was planning to sail back to Rome with Cleopatra and her children in chains. Her worst fears confirmed, the pharaoh prepared to take her own life.

There is no question that Cleopatra died at this time, but the exact manner of her death remains a mystery unresolved and still eagerly debated over 2,000 years later. The popular version, as propagated by Shakespeare's *Antony and Cleopatra*, has Cleopatra clasping two asps to her breast and dying soon afterwards from their poisonous bites. Roman writer Suetonius says that Cleopatra's handmaidens were also bitten by the snakes and then arranged Cleopatra's crown and robes before they too succumbed to the poison.

Plutarch's account then has Octavian bursting into the mausoleum as Cleopatra's handmaidens finish their final duties and demanding angrily: "Was this well done of your lady?" The reply from one of the handmaidens, who was scarcely able to hold up her head, was "extremely well done, as befitting the descendants of so many kings". She then followed her mistress in death. Octavian, soon to pronounce himself Augustus, *princeps* (first citizen) of Rome, would be denied the glorious spectacle of parading Cleopatra in a triumph, the ritual humiliation that befell many foreign leaders defeated by Rome.

A POISONED HISTORY

Did Cleopatra kill herself with an asp? It is a question still debated today. Her suicide was described in contemporary Roman accounts by Plutarch, Dio, Suetonius and Pliny and, more recently, by Shakespeare. Today, it is thought the name "asp" actually refers to the Egyptian cobra (*Naja haje*). The cobra is an ancient symbol associated with the Egyptian goddess Isis and worn on the pharaoh's headdress. Varying accounts state that the snakes were brought to Cleopatra in a basket of figs, the figs themselves were poisoned, or that Cleopatra was found with two puncture marks in her arm, through which the toxin was delivered by poison comb or brooch pin. Yet another has Cleopatra smearing a poison ointment on her skin or biting her own arm and rubbing the poison into the wound. A recent study has rejected the idea of death by cobra. A cobra bite is not always fatal, but when it is, the death is painful and long, and leaves its victim in a state of paralysis, with the possibility of contorted facial features and wide eyes. This, the study argues, would not leave the queen in a final state of peaceful bliss. It also contradicts Plutarch's suggestion that Cleopatra had found an asp bite to be a painless death akin to a deep sleep. Cleopatra's interest in poisons would surely have led her to use one that delivered a peaceful death. A popular theory today is that Cleopatra ingested a toxic cocktail of opium, aconite and hemlock that allowed her to slip away. Others suggest Octavian simply murdered Cleopatra and then invented the story of the snake suicide as it fitted in neatly with the asp symbol from Egyptian mythology. It remains an unresolved question for the ages.

ABOVE: Jean André Rixens' 1874 The Death of Cleopatra *propagated the myth of the pharaoh as a peaceful, beautiful corpse.*

Mithridates' Mass Suicide

King Mithridates of Pontus had good reason to fear poisoning. His father was poisoned at a banquet, his mother plotted to assassinate him, and he was an enemy of the great poisoners of the ancient world: the Romans.

To fortify himself against these threats, the teenaged Mithridates left court and became a hardened survivalist in the remote wilderness. He also took homeopathic doses of different poisons, to develop a resistance to as many as possible.

In 115 BC Mithridates returned from the wilderness to the capital of Pontus, in northern Anatolia, and began a purge of the royal court. He had his mother executed and his brother arrested, and proclaimed himself king. He promised to free the people and to extend the royal borders tenfold. He made a convincing case. After his exile in the wild, Mithridates had grown into an energetic and physically commanding figure. He was also, according to Roman historian Pliny the Elder, a "brilliant intellect … an especially diligent student of medicine [who] collected detailed knowledge from all his subjects".

Behind Mithridates' pursuit of knowledge was self-preservation: he gathered doctors, scientists and Scythian shamans to help create a universal remedy against any known poison. He also cultivated his own poisons with his botanist Krateuas and corresponded about popular antidotes with Zopyrus, personal physician to the Egyptian pharaoh Ptolemy.

Mithridates experimented on condemned criminals with poisons and possible antidotes. For himself, the king devised a potion he named *Mithridatium*: a pill made up of a paste of different drugs, medicines and poisons and bound together with honey. According to Pliny, Mithridates then "thought out the plan of drinking poison daily, after first taking remedies, in order that the sheer custom might render it harmless."

While his war against assassination went on, Mithridates waged an expansionist campaign against his neighbours. He defeated the Scythians to the north and the Sarmatians to the west and made an alliance through marriage in Armenia. But when he brokered the Armenian invasion of Roman ally Cappadocia, Rome could no longer ignore Mithridates' ambitions to create an empire to match those of his proclaimed forbears, Darius I of Persia and Alexander the Great.

After Rome tried to invade Pontus, Mithridates responded by massacring over 80,000 Roman citizens as he conquered and razed Roman cities in Anatolia. Rome went after Mithridates, but even the formidable general Sulla could not defeat the king in war and was forced to parley. As Sulla's legionaries retreated, Mithridates formed an army large enough to take on the might of the Roman republic; the conflict would consume him and his kingdom.

A second and third Mithridatic War with Rome followed. The republic won some early victories, but Mithridates pushed back with equal vigour. He sacked the sacred shrine at Delphi, before allying himself with the Cilician pirates, the Ptolemaic Dynasty of Egypt, and briefly the rebel slave leader Spartacus. For a time it looked as though Mithridates could unite the eastern and western cultures of Asia minor into a Hellenistic empire to rival Rome, but he had not counted on family betrayal and the cunning of Roman general Gnaeus Pompey. Pompey crushed Mithridates' allies, including the Cilician pirates, and befriended any neighbours who would betray him – even his own son, Machares.

In response, Mithridates murdered Machares, but not before his other son Pharnaces joined with

OPPOSITE: A bust of Mithridates wearing a lion's head cap.

Rome. Pompey now began closing in. Mithridates' last stand came in one of his own citadels as Roman legionaries marched towards him. Trapped with his two young daughters, Mithridates decided to take his own life rather than suffer the indignity of being paraded in chains in a Roman triumph. According to Appian's *Roman History*, Mithridates then took out some poison that he carried next to his sword belt and mixed it. But his daughters insisted they be allowed to drink the poison first:

"The drug took effect on them at once; but upon Mithridates, although he walked around rapidly to hasten its action, it had no effect, because he had accustomed himself to other drugs by continually trying them as a means of protection against poisoners. These are still called the Mithridatic drugs."

It seemed that Mithridates' lifetime of taking homeopathic doses of poison had made him immune to the very poison that he intended to end his life. But he would not be denied his suicide: he asked a Gaulish bodyguard to run him through with a sword and died where he fell. In that same year, 63 BC, Pompey annexed Pontus into the Roman republic and put to death Mithridates' remaining children, wives, mistresses and sisters. The royal line was finished.

THE MITHRIDATE ANTIDOTE

The recipe for *Mithridatium*, also known as the Mithridate Antidote, was reportedly found in the personal cabinet of King Mithridates and delivered to general Pompey. According to Pliny the Elder, *Mithridatium* contained over 54 different ingredients: these were variously said to include dried walnuts, figs and leaves of rue, and opium, myrrh and castoreum, a substance found in beaver's testicles. Pliny also mentions that Mithridates drank the blood of ducks that had fed on poisonous plants. Later, Nero's physician Andromachus and Marcus Aurelius' physician Galen both created updated versions of *Mithridatium* to prevent the emperors being poisoned, a common occurrence among Roman patricians. One antidote contained 57 ingredients including the flesh of a viper, and Nero was said to have taken it every day. Recipes for universal antidotes went on to become popular among scholars in Islamic countries as well as re-emerging in the West. Some allegedly protected people from the plague in medieval Europe, and by the time of the Renaissance the antidote industry was booming. "Venetian treacle" was one popular antidote, sold in ornate glass jars; less expensive versions in cheaper vessels were sold in poorer apothecaries. If the antidotes did not work, then the apothecary was blamed for using sub-standard ingredients. Later, city officials supervised the preparation of antidotes, a practice that eventually became the regulation of modern medicine. Black market versions of antidotes grew alongside the official ones. Standardized recipes of antidotes were made easily available through the invention of the printing press. In 1618 the official formula for *Mithridatium* was published in the *London Pharmacopoeia*, although scepticism about the benefits of drugs with multiple ingredients was growing. By the nineteenth century, *Mithridatium* had fallen out of favour altogether.

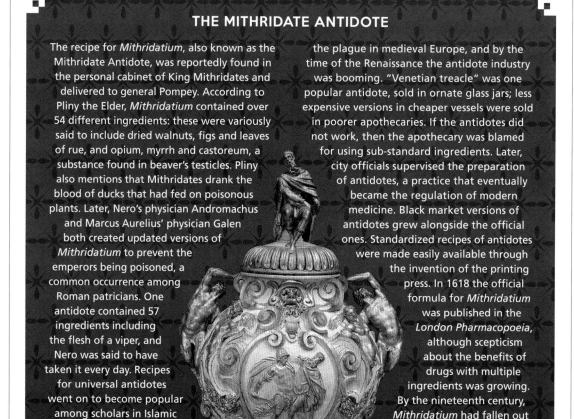

LEFT: A Mithridatium *container*.

ABOVE: As Roman legionaries storm the citadel, Mithridates breathes his last. Immune to his own poison, the king's bodyguard delivered the coup de grâce.

Belladonna

Atropa belladonna is a member of the nightshade family (Solanaceae) and is commonly known as deadly nightshade. It is also called dwale, death cherries and devil's herb.

OVERVIEW:
Belladonna is a tall, bushy plant which produces small, shiny black berries that are often mistaken for fruit. The plant contains the toxic alkaloids hyoscyamine, scopolamine and atropine, and is highly poisonous. Even light contact with belladonna leaves can cause a rash on human skin. The leaves, roots and berries are all used in the preparation of belladonna poison, which early hunters and gatherers added to the tips of their spears and arrows. During the Middle Ages, witches rubbed a concoction of belladonna and bear's grease into their skin to produce hallucinatory sensations of flying. Popular during the Renaissance, belladonna is believed to be the poison ingested by Juliet to feign death in Shakespeare's *Romeo and Juliet*. In Italy during this time, fashionable women used belladonna eye drops to dilate their pupils and give themselves a "doe-eyed" appearance. This is where the poison gets its name, belladonna, or "beautiful woman" in Italian.

TOXIC EFFECTS:
The main poison in belladonna is atropine, a powerful anticholinergic that blocks the action of acetylcholine in the human body. Acetylcholine is a neurotransmitter which sends signals between the central nervous system and cells. Atropine specifically affects the part of the nervous system that controls the heart and digestive system. In small medicinal doses, belladonna is therefore used to treat stomach spasms and speed up the heart. However, in larger doses it can quickly become lethal. Even one berry from the belladonna plant can be fatal, although people have survived swallowing over one whole gram of the berries.

SYMPTOMS:
The symptoms of belladonna poisoning set in within minutes and can last for over 10 hours, or even several days in serious cases. Early symptoms include a dry mouth and throat, dilation of the pupils, nausea and vomiting. If left untreated, victims can experience hallucinations, delirium, a staggering gait and extreme drowsiness. In fatal cases, the victim's face becomes flushed, their breathing jerky and their extremities grow cold. A coma and a rapid, intermittent pulse immediately precede death.

TREATMENT:
The antidote for belladonna poisoning is the drug Physostigmine, which acts by increasing the concentration of acetylcholine in the body's nervous system.

FAMOUS POISONINGS:
- Nero murdered his half-brother Britannicus in AD 55 using a belladonna tincture prescribed by the professional poisoner, Locusta (see pages 33–34).
- Swiss nurse Marie Jeanneret murdered seven of her patients in 1868 after becoming obsessed with belladonna and even experimenting with it on herself.
- Eleventh-century Scottish king, Duncan I, killed an army of invading Danes led by Viking Svein Knutsson by feeding them belladonna-infused liquor.

OPPOSITE: The Atropa belladonna *plant.*

Agrippina and Nero: A Family Affair

One of ancient Rome's most notorious poisoners was Agrippina, wife to emperor Claudius and mother to emperor Nero. Beautiful and deadly, Agrippina had seduced and married Claudius after removing all rivals who stood in her way and allegedly poisoning her previous husband.

Once Empress of Rome, Agrippina prepared the ground to make Nero emperor. Her method for this was poison.

Agrippina had little love for Claudius, but the emperor had to move mountains to marry her. To do this he executed his previous wife Messalina and changed the law to make incestuous unions legal; Agrippina was Claudius' niece. Once married, Agrippina removed anyone who was a threat to her position, especially those who had been loyal to Messalina. But unfortunately for Agrippina, Messalina had provided Claudius with a son, Britannicus, and he was the heir to the throne.

After some haranguing, Agrippina convinced Claudius to adopt Nero (then called Lucius) as his heir. But Claudius later reneged on the deal and even repented for marrying Agrippina in the first place. When he started preparing Britannicus for the throne, Agrippina called upon her favourite poisoner, a Gaulish woman called Locusta. Locusta was in jail at that time for poisoning, and agreed to help assassinate Claudius in return for her freedom.

The plan was kept simple: Agrippina would make Claudius drunk so he was off his guard. She would then feed him a meal of mushrooms, his favourite food. In the meantime, Locusta would make sure Claudius' assistant and food taster was indisposed. Because Claudius would be drunk by this time, he would let the disappearance of his food taster pass.

The plan ran like clockwork. Claudius wolfed down his mushrooms and then began clutching his stomach in pain. Soon he was experiencing cramps and cold sweats; before long, he was writhing on the floor and gasping for air. As these are the common symptoms of poisoning by fly agaric (page 11), it is possible Claudius' mushrooms were laced with the poison toadstool. A physician was quickly summoned and called for a feather to be stuck down Claudius' throat to induce vomiting.

This was done, but Locusta was one step ahead. The feather provided had been soaked in poison ahead of time. The mushrooms themselves had only been a ruse to make Claudius sick; it was in fact the feather that delivered the killing blow. The plan worked with lethal precision; Claudius was dead. Nero was now made emperor, but there was no confusion about the manner of his succession. Senators reportedly sniggered openly as Nero read out Claudius' eulogy.

If Locusta had expected to receive praise and reward for successfully poisoning Claudius, she was wrong. Instead, Agrippina made Locusta the scapegoat for the murder and had her sentenced to execution. However, Locusta's efforts had not gone unnoticed by the new emperor, who still had a rival in his step-brother Britannicus. Worse still, Agrippina had disapproved so strongly of an affair Nero was having with plebeian woman Claudia Acte that she began to support Britannicus' claim to the throne. The simplest thing, Nero decided, was to

OPPOSITE: In Joseph-Noel Sylvestre's painting, Locusta tests poison on a slave.

avail himself of Locusta's services and do away with Britannicus so any threat to his rule was removed.

Once freed from prison, Locusta came up with a plan for poisoning Britannicus at a dinner party at which Nero and Agrippina were both present. At the party, Britannicus was brought some wine mixed with hot water, as was often the custom. Britannicus' taster tested the wine, but not the cold water that Britannicus added to make the drink cool. And this was where Locusta's poison lay.

The contemporary historian Tacitus recorded that as Britannicus writhed on the floor and clutched his throat, some guests quickly rose and left the party, and others sat frozen as they looked at Nero. Nero, in response, calmly stated that this was a normal incident and that Britannicus was simply having an epileptic fit. The party, he countered, should resume. But for Nero's mother this was a moment of pure horror, as Tacitus reports:

"From Agrippina, in spite of her control over her features, came a flash of such terror and mental anguish that it was obvious she had been as completely in the dark as the prince's sister Octavia. She saw, in fact, that her last hope had been taken – that the precedent for matricide had been set."

At Nero's behest, the dinner party continued, but Britannicus, who was taken to his quarters, succumbed to the poison a few hours later. Nero's message, as understood by Agrippina, was clear. Nobody was safe from the emperor. And now, with Britannicus gone, Nero could turn his murderous attentions to other members of his family: including his mother.

First Nero stripped Agrippina of all her honours, privileges and powers and exiled her from the royal palace. He then asked Locusta to poison Agrippina during various dinner parties; the plan failed three times because Agrippina had ingested a series of antidotes before arriving. After one dinner, Nero loaned Agrippina a sabotaged boat, which collapsed on her way home across the Bay of Naples; Agrippina, however, was able to swim to shore.

In the end, Nero had his mother stabbed to death; he reportedly then stood over her corpse and commented on its good and bad features. Tacitus reports that astrologers had warned Agrippina that her son would become emperor and kill her, to which she replied: "Let him kill me, provided that he becomes emperor."

OPPOSITE: A bust of Agrippina, the mother Nero murdered.

LOCUSTA

According to contemporary annals by Roman writers Suetonius, Tacitus and Cassius Dio, Locusta had learned her trade in poisoning in her home country of Gaul before moving to Rome. Locusta was said to favour the plant belladonna, although she almost certainly had a working knowledge of others. Her expertise was such that Nero helped her set up a poisoning school after she helped him assassinate Britannicus. She was also rewarded with land and money, before Nero's successor Galba executed her in 69 AD. However, Locusta was not the only female poisoner of ancient Rome: Canidia and Martina were two others.

According to the writer Horace, Canidia specialized in administering honey laced with hemlock and was strong enough to tear apart a lamb with her teeth. Martina was alleged to have poisoned Germanicus, nephew and adopted son of Tiberius. If he had not been assassinated, Germanicus would have been made emperor; instead, the throne went to his son Caligula. To evade capture for the crime, Martina killed herself with a poison found in a vial knotted in her hair; no one is sure who ordered the killing of Germanicus.

Aconite

Aconite is a poisonous alkaloid and the active component found in over 100 species of plant belonging to the genus Aconitum. *The most common of these plants is* Aconitum napellus, *which has purple hood-shaped flowers and thick tuberous roots, and is known by the names monkshood, wolfsbane and bear's foot*

OVERVIEW:

Aconite is known as the "Queen of Poisons", and every part of this plant is highly toxic. According to mythology, the plant comes from the hill of Aconitus, where Hercules fought Hades' three-headed guard-dog, Cerberus. Saliva from Cerberus was said to have landed on the plant, making it poisonous. In the medieval age, witches used aconite along with belladonna and henbane as an ointment to induce the sensation of flying (page 10). The plant at this time was also feared, because people believed witches might use it to summon the devil. Shakespeare refers to aconite in two of his plays: Romeo takes the poison to commit suicide in *Romeo and Juliet*, and the "tooth of wolf" used by the witches in *Macbeth* is aconite going by its alternative name, wolfsbane. Arrows dipped in aconite were fired at wolves during the Middle Ages to kill them, so becoming a "wolf's bane". Aconite's tuber-shaped root is traditionally used, after careful preparation, in Chinese and Hindu medicine to treat rheumatism, neuralgia, lumbago and cardiac issues. However, one teaspoon of the unprepared root is lethal to a human adult, and even handling the plant is dangerous.

TOXIC EFFECTS:

Aconite poisoning causes the heart to beat abnormally by sabotaging the pumps that move sodium ions in and out of the heart's cells. By binding to these pumps, the poison holds them permanently in an "on" position. Without treatment, the heart will contract and beat abnormally. This is followed by asystole, the most serious form of cardiac arrest, which is usually irreversible. Nerve and muscle cells that rely on sodium channels are often affected too, which results in muscle spasms and seizures. Death from an aconite overdose can occur any time after ingestion, from a few minutes to a few days. Just a couple of milligrams of aconite, weighing as much as a sesame seed, can kill an adult.

SYMPTOMS:

The symptoms of aconite poisoning occur rapidly and include a burning or tingling sensation on the lips and throat, followed by excessive salivation, nausea, vomiting, speech impairment, numbness in the throat, difficulty breathing and visual disturbances. Dizziness, muscle weakness, hallucinations and sensory disturbances follow. Death is normally preceded by a massive cardiac arrest.

TREATMENT:

There is no specific antidote for aconite poisoning, so treatment is mainly supportive. Atropine can be given to treat a low heart rate and activated charcoal to decontaminate the intestines. Various other drugs used to stabilize the heart rate can be administered, and heart bypasses are sometimes performed in extreme cases.

FAMOUS POISONINGS:

- In 2009, a British woman poisoned her lover's curry with aconite to prevent him marrying a younger rival. Lakhvinder "Lucky" Cheema, 39, died hours after eating the curry laced by Lakhvir Singh, 45. It was the first aconite poisoning in the UK since 1882.
- In 1882, Dr George Lamson was hanged for poisoning his 18-year-old brother-in-law Percy John for his inheritance. Lamson took a Dundee cake to Percy's school that contained raisins laced with aconite.

OPPOSITE: *The Aconitum napellus plant.*

Lead

Lead is soft grey metal that occurs naturally in the Earth's crust, but is also produced by human activity, such as mining, burning fossil fuels and manufacturing. It is a useful metal because it is soft enough at room temperature to bend or work into different shapes.

OVERVIEW:
Lead has been used in pipework since ancient times, and was favoured by the Romans not only in their plumbing but also in their utensils. Roman plates and cups were often made from lead, and lead compounds were also used in their make-up, wall paint, hair dye, medicine and tooth fillings. The Romans even used "sugar of lead" to sweeten their wine. This abundance of lead has often posed the question of widespread lead poisoning throughout the Roman Empire, especially among the patrician class. These aristocrats were known to have been affected by low fertility rates and chronic illnesses that are consistent with lead poisoning. In modern times, lead poisoning has continued to be a threat, especially among children who can ingest flakes of lead-based paint, which was used until the late 1970s. Lead poisoning is particularly harmful to children as it can hinder the chemicals enabling bone and brain growth, and seriously impair their neurological system.

TOXIC EFFECTS:
Lead poisoning is often a chronic problem, as small amounts can build up in the body over time without producing detectable effects. Lead poisoning is therefore cumulative, as it remains in the bones for as long as 32 years and in the kidneys for seven. Chronic lead poisoning can come from a variety of sources, such as paint, vessels that contain lead, old batteries, roofing materials, or simply from contaminated air, soil or food. The toxic effects of lead over time can lead to irreversible brain damage, especially in children. In higher doses, lead can cause serious damage to the kidneys and nervous system, leading to seizures, unconsciousness and death.

SYMPTOMS:
Early symptoms of chronic lead poisoning present themselves as irritability, loss of appetite, constipation, depression, abdominal pain, fatigue, decreased libido, insomnia, memory loss, and weakness and tingling in the extremities, followed by the appearance of a bluish line on the gums. Acute symptoms include vomiting, diarrhoea, a metallic taste in the mouth, shock, and decreased urination accompanied by muscle weakness, pain, tingling, headache, convulsions, coma and death.

TREATMENT:
A simple blood test can determine the level of lead in a victim's body, followed by the administration of chelating agents to purge the lead from the body via the urine. Stomach flushing is often used in cases of severe, acute poisoning.

FAMOUS POISONINGS:
- Lead found in Beethoven's bones and hair probably caused his untimely death at the age of 56 in 1827. Beethoven was sick for much of his life, was easily angered and suffered from depression, all of which can be attributed to lead poisoning. He consulted many doctors during his lifetime, but none were able to help him. He insisted his body be tested for poisons after his death, so others would not suffer as he did.

OPPOSITE: A lead pipe from Ancient Rome.

秦始皇

姓嬴名政始皇目始皇乙卯即王位庚辰併天下稱皇帝
在位三十七年居王位二十五年即帝位十二年壽五十

Purging Qin Shi Huang

When Qin Shi Huang crowned himself the first emperor of China in 220 BC, he boasted that his dynasty would last 10,000 generations. He then demanded that an elixir of immortality be brought to him so he could see this prediction through.

But rather than provide him with eternal life, the pills and potions that Qin ingested daily almost certainly killed him.

Born Ying Zheng in 259 BC, Qin Shi Huang is best known for providing some of the great icons of Chinese history. He unified the country, constructed connecting roads and canals, and linked together its fortresses to create the Great Wall of China. The emporer then built a vast army to expand his borders, while enslaving and castrating local populations as he went. There were said to be many eunuchs at Qin's court.

To make sure future generations would remember the might of his formidable fighting machine, Qin ordered that each soldier be recreated in terracotta. This 8,000-strong terracotta army was later buried next to Qin's mausoleum and left undisturbed until 1974, when astonished farmers uncovered it.

Qin himself was described as a "man of scant mercy", with a "puffed-out chest like a hawk". He ruled as a tyrant and did not hesitate to use murder to remove anyone he considered a threat. He buried over 400 Confucian scholars alive and ordered the wholesale burning of books. History, Qin said, was irrelevant, although it has been suggested that his attacks on scholars were to concentrate their minds on discovering eternal life. Ancient Chinese historian Sima Qian said that if "Qin should ever get his way with the world, then the whole world will end up his prisoner".

Qin spent a fortune constructing his vast mausoleum and issued an executive order that the elixir of immortality be brought to him. Scared officials scouring the land for Qin would write awkward messages on wooden slats that, so far, they had had no success. One official said that maybe a herb from a local mountain could fit the bill. Many searchers simply did not return, realizing that the cost of failure was execution.

Meanwhile, Qin set hundreds of alchemists to work on creating an elixir; many of these substances contained dangerous substances such as mercury and arsenic. The search by alchemists for elixirs to prolong life would continue in China for over 2,000 years after Qin died. Qin himself took a daily dose of his alchemists' elixir and died suddenly in 210, almost certainly of mercury poisoning. It is said that tiny rivers of liquid mercury flow through Qin's lavish inner mausoleum, which has been left unexcavated.

OPPOSITE: Emperor Qin Shi Huang.

RIGHT: Qin's unearthed terracotta army.

Mercury

Mercury is a metallic element, found in the Earth's crust, which carries the chemical symbol Hg. It is the only metal that remains a liquid at room temperature.

OVERVIEW:
Mercury is an extremely toxic poison that is almost impossible to avoid. There is mercury in our food, air and water, and some of us have mercury in our amalgam fillings. It is thought that an average daily intake of mercury for adults is around 3 mg and at any time there is around 6 mg in the body. While this dosage is not enough to do us permanent damage, people consumed this amount or more in various medicines from the last century, usually to treat embarrassing conditions such as constipation or syphilis. Ancient rulers consumed liquid mercury as elixirs, and evidence of its use from antiquity have been found in India, China and Egypt; mercury traces have also been discovered in ancient Egyptian tombs. More recently, nineteenth-century milliners used mercury to make hats and often suffered from mercury poisoning; the Mad Hatter from Lewis Carroll's *Alice in Wonderland* (pictured) was most probably based on these sufferers.

TOXIC EFFECTS:
Mercury's toxicity depends on its form. Liquid mercury can be reasonably tolerated in low doses as it is then expelled by the body, rather than being absorbed by the intestinal tract. For this reason, mercury poisoning can occur over a long period if ingested. Mercury vapour, by comparison, is a fast-acting poison that attacks the human body through the lungs. Acute mercury poisoning quickly damages the respiratory system, internal organs and central nervous system, resulting in brain, kidney and lung failure, pneumonia, psychosis and death.

SYMPTOMS:
Initial symptoms of mercury poisoning can include anxiety, loss of memory and depression, alongside muscle weakness, nausea, numbness in hands and feet, and difficulty breathing and walking, followed by vomiting, shaking, impaired motor skills, loss of hand-eye coordination and disorientation. Symptoms of long-term mercury poisoning often appear in the mouth, where the gums become covered with a grey film and teeth and gums deteriorate. Another chronic symptom is the excessive flow of urine followed by an inability to pass water, a sign of severe kidney damage.

TREATMENT:
The drug dimercaprol is administered via an injection as an antidote for acute mercury poisoning. Dimercaprol works by binding to heavy metals, as does chelation therapy, which helps to transport mercury from the organs and into the bloodstream so it can be eliminated in a victim's urine.

FAMOUS POISONINGS:
- Several sailors aboard the HMS *Triumph* and HMS *Phipps* were taken ill in 1810 after a load of mercury was salvaged from a wrecked Spanish ship. Their symptoms included tooth loss, skin disorders, excessive salivation and paralysis.
- American professor of chemistry Karen Wetterhahn died in 1997 from mercury poisoning after researching the metal for many years. It was found that the gloves Wetterhahn used while studying mercury were not sufficient to prevent it being absorbed through her skin.
- In 2008, American Tony Winnett died from mercury vapours after trying to extract gold from computer motherboards. His home had to be gutted because the mercury contamination was so severe.

Temptress Empress Wu

The life of Empress Wu Zetian, who ruled medieval China for over 50 years, remains controversial. Once a low-ranked concubine of the royal court, Wu is remembered as an enchantress and usurper who used sex and poison to gain power.

According to one Chinese history text, "She killed her sister, butchered her elder brothers, murdered the ruler, poisoned her mother. She is hated by gods and men alike".

Wu Zetian was born in AD 624 to a rich family and as a young woman became a fifth-ranked concubine at Emperor Taizong's royal court. Wu was responsible for changing the imperial sheets and was said to indulge the emperor in some of his more unusual sexual appetites. After Taizong died, Wu was able to take up with his son and heir Gaozong.

According to contemporary histories, Wu killed the daughter she had with Gaozong and framed his wife, the empress Wang, for the murder. Gaozong believed Wu's story and on her advice jailed Wang along with his top concubine. Wu subsequently became empress consort and ordered the women's hands and feet to be cut off and their bodies thrown into a vat of wine. "Now these two witches can get drunk to their bones," Wu reportedly said.

Wu then did away with all of those who had resisted her promotion to empress consort: she had them executed, demoted, exiled, or forced to commit suicide. When a cousin of Wu's was made Gaozong's concubine, the empress poisoned her and then had her own brothers executed for the crime. As Wu continued removing opposition from the royal court, Emperor Gaozong began experiencing headaches and a mysterious illness, which ended his life in 683. Unusually for a Chinese emperor, Gaozong died alone, with only Wu at his side.

Wu had already poisoned the crown prince Li Hong, her son, and exiled other potential heirs who stood in her way. Although her other sons were the official emperors for a period, Wu was the real power behind the throne until she took full control in 690. She then ruled with an iron fist, forming a secret police to punish detractors and encouraging courtiers to spy on others and report their wrongdoings.

Many today argue that Wu's reign was unfairly judged by contemporary historians because she was a woman; emperors behaving the same way would not have been so vilified. In reality, Wu was a highly successful empress who steered China through one of its golden periods. However, her giant funeral tablet received no epithet and today remains blank. For many, Wu will always be the temptress who poisoned her rivals and used sex to gain control.

ABOVE: Wu is depicted with later general, Yue Fei.

OPPOSITE: A Tang dynasty print of Wu Zetian.

Chapter 2: Medieval and Renaissance Poisons

Poison became a popular form of killing during the Middle Ages. At this time, European monks wrote about the poisons of plants and animals, often by plagiarizing texts from the ancient Greeks and Romans.

These texts were of little use to the largely illiterate population, who instead sought their cures from village healers, or witches, or through prayer, using talismans and amulets to ward off unnatural maladies or supernatural interference. Meanwhile in the East, Arab alchemists developed one of history's greatest poisons: arsenic. Through the process of extracting, distilling and crystallizing arsenic into a powdered form, these alchemists not only made the poison colourless and odourless, but also helped invent the science of chemistry.

Because powdered arsenic was difficult to detect, it became a favourite among the poisoners of Renaissance Europe. Poison was in particular vogue in Italy at this time, and physicians, alchemists and pharmacists would prepare poisons for aristocratic families from the "the three kingdoms of nature": animal, plant and mineral.

Arsenic made up the base for the infamous poison La Cantarella, administered by Pope Alexander VI's family, the Borgias. Other poisons mentioned in Italian Renaissance texts included snake venom, cantharidin, aconite, belladonna and strychnine; some of these were used by freelance assassins employed by the Venetian Council of Ten. "The Ten" was one of the city's governing bodies; their clandestine activities included murdering those considered a threat to the state. In 1450, the Council sanctioned the murder of Count Francesco Sforza, Duke of Milan, by placing poisonous balls in his fireplace. These, supposedly, would emit toxic fumes that would kill any who sat near it. The plot, however, did not succeed.

As poisoning reached its zenith in Italy, it spread to other parts of Europe. Italian aristocrat Catherine de' Medici was accused of single-handedly introducing the fashion of poisoning to France, alongside high-heeled shoes and lavish court festivals called "magnificences". De' Medici tested her poisons on the poor and sick to observe their effects, but during the Renaissance a more scientific approach to poisons also developed.

The ubiquity of poisoning in the Renaissance meant that it became of great interest to physicians, particularly the Swiss physician Paracelsus. Paracelsus raged against the prescribed wisdom of ancient physicians and demanded that modern doctors had a working knowledge of chemistry. He also famously noted that the dosage of a specific chemical determined the difference between its therapeutic and toxic properties: "What is there that is not poison? All things are poison and nothing is without poison," he said.

For ushering in a new era of chemical medicine, Paracelsus is often known as the father of toxicology. Scientific treatises on toxins and their effects followed, but it would be some time before poisoning became solely the domain of science, as opposed to the malevolent, supernatural world of witches, sorcerers and black magic.

ABOVE: This sixteenth-century woodcut shows Paracelsus performing an operation on a patient's head.

The Bloated Borgias

One August night in Rome in 1503, Cesare and Rodrigo Borgia were gripped by the effects of a violent poison. Cesare lay in bed, his face an angry purple and skin peeling.

Meanwhile Rodrigo, better known as Pope Alexander VI, succumbed to the poison. The next day his body was exhibited to the people of Rome. It was reported to be "the ugliest, most monstrous dead body that was ever seen."

Few mourners could bear to gaze upon the increasingly foul face of the dead pope in state, which turned the colour of mulberry, was covered in blue-black spots and was "far more horrifying than anything that had ever been seen or reported before".

Renaissance historian Raffaello Maffei described the spectacle: "It was a revolting scene to look at that deformed, blackened corpse, prodigiously swelled and exhaling an infectious smell; his lips and nose were covered with brown drivel, his mouth was opened very widely, and his tongue, inflated by poison, fell out upon his chin; therefore no fanatic or devotee dared to kiss his feet or hands, as custom would have required."

Horror aside, the greatest surprise about Rodrigo's death was that he was usually the poisoner, not the poisoned. Murder was only one of very many crimes linked with the House of Borgia during Alexander's reign: adultery, incest, rape, simony, theft and bribery were some of the others. The Borgias' reputation as murderers was legend; they killed not only to further their ambitions but sometimes simply out of malice.

Originally from Spain, the Borgias worked their way up through the political and clerical elite in fifteenth-century Italy and rivalled other leading families such as the Sforza and the Medici. Alfonso was the first Borgia to become Pope in 1455. He immediately showered favours on his nephew Rodrigo, making him a cardinal and grooming him for the papacy. Rodrigo himself reached this position in 1492. However, his reign was so corrupt and devoid of spiritual significance that it helped lead directly to the revolutionary changes of the Protestant Reformation.

After being made pope, Rodrigo made his teenaged son Cesare a cardinal and lavished clerical titles on other relations: he wanted to create a lasting dynasty. Further alliances were made through the marriages of Rodrigo's daughter Lucrezia, a hazel-eyed beauty whose face in profile inspired many portrait painters. Lucrezia was also something of a femme fatale, said to poison those whom she could not seduce and to have incestuous relations with family members, including her brother Cesare.

LA CANTARELLA

There are no surviving recipes for the poison known as La Cantarella, but it is thought to have been a toxic mix of phosphorus, lead acetate and arsenic. Some accounts add cantharidin to the mix. One description of manufacturing La Cantarella has the poisoner sprinkling arsenic on the entrails of disembowelled pigs and letting it fester. The entrails would then be wrung out and the putrid mixture left to dry before being converted to a powder.

OPPOSITE: Lucrezia Borgia.

Lucrezia was said to attend orgies held by Cesare and Rodrigo in the Vatican palace. The most famous of these was the Banquet of Chestnuts, a dinner party with 50 prostitutes as guests.

When not seeking earthly pleasures, Rodrigo and Cesare worked to consolidate their rule and vast wealth. At the head of a papal army, Cesare brought the northern Italian states under his control and formed an alliance with Ferdinand and Isabella of Spain. Those who could not be bought or conquered were murdered. Often the extremely toxic arsenic-based poison called La Cantarella was used for the task.

Each Borgia family member was reported to have a different method for administering La Cantarella to a foe. Cesare had a lion's head ring with two sharp canines on the bottom that could prick his victim's hand with the poison. Rodrigo also administered the poison by pricking: he would invite his intended victim into his papal chambers and then ask if they could help him unlock a cabinet with a key that was being difficult. However, the key was fashioned with a pin that would prick them once they took it.

Lucrezia was said to have a ring with a hinged lid so that La Cantarella could be poured into a victim's wine. The poison was thought to have a slightly sweet taste, so would be ingested unnoticed. Attending a dinner party at the Borgias was often a life or death matter, but in the end it was a fatal error in planning that led to Rodrigo's death.

Rodrigo and Cesare had both been invited to dinner at the house of Adriano Castellesi – a cardinal they planned to murder – on a hot night in 1503. The Borgias had brought with them gifts of wine, including a poisoned bottle that Rodrigo mistakenly began serving to himself. Seeing this and believing the wine to be from an unadulterated bottle, Cesare also partook. The accident killed Rodrigo, but Cesare survived. However, he failed to become Pope and died in 1507 in Spain.

The violent and scandalous history of the Borgias has often been disputed by modern historians, who say there is little evidence to back up the claims. Instead, Lucrezia has recently been recast as a pawn in her father's ambitions and the Borgias' legacy as poisoners said to be exaggerated. However, Lucrezia, Cesare and their father Rodrigo will always be remembered as poisoners. One contemporary cleric described Rodrigo as having "given his soul and body to the great demon in Hell".

And what of the poison that apparently so disfigured Rodrigo as he lay in state? A modern theory suggests that he died instead of malaria. A conflicting contemporary account says it was not poison that made his face an unnatural mulberry colour, but simply rapid decomposition caused by the sticky Roman summer heat.

OPPOSITE: *John Collier's 1893* A Glass of Wine with Caesar Borgia *depicts a potentially dangerous dinner date.*

THE BANQUET OF CHESTNUTS

The Borgia orgy known as the Banquet of Chestnuts was described by contemporary papal chronicler Johann Burchard: "On the evening of the last day of October, 1501, Cesare Borgia arranged a banquet in his chambers in the Vatican with 'fifty honest prostitutes' … who danced after dinner with the attendants and others who were present, at first in their garments, then naked. After dinner the candelabra with the burning candles were taken from the tables and placed on the floor, and chestnuts were strewn around, which the naked courtesans picked up, creeping on hands and knees between the chandeliers, while the Pope, Cesare, and his sister Lucrezia looked on. Finally, prizes were announced for those who could perform the act most often with the courtesans."

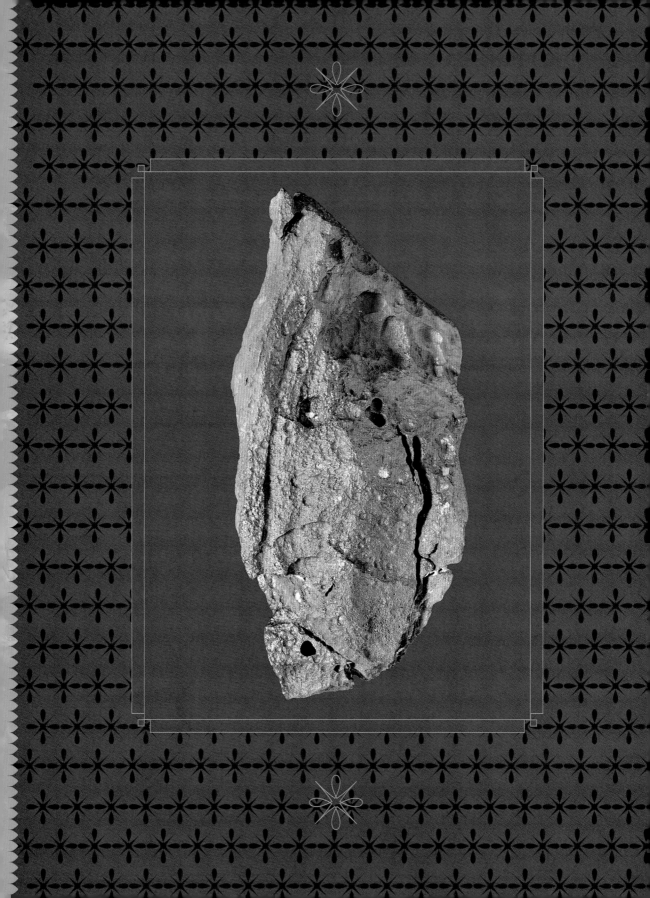

Arsenic

Arsenic is a grey metallic element found in the Earth's crust. It also exists in trace amounts in all human tissue. As a poisonous compound, arsenic commonly appears as the sugar-like powder white arsenic (arsenic trioxide).

OVERVIEW:
Arsenic was not isolated in its elemental form until the Middle Ages, but its ores – realgar and orpiment – were known to ancient civilizations such as the Assyrians, Chinese and Romans. The Romans, in particular, knew how to make white arsenic, and used it in pesticides, medicines and poisons alike. When added to food or drink, white arsenic does not change colour or give off an odour, but instead produces a slightly sweet taste. Arsenic added to wine was a favourite way to assassinate people in Renaissance Europe. Arsenic in small amounts has been prescribed as a medicine since ancient times and was the leading ingredient in Fowler's Solution, a medicine of the nineteenth century. The solution was variously given for ailments that included epilepsy, skin conditions and syphilis. Arsenic was also used as a pigment since ancient times to create colours such as red and yellow, and, from the late eighteenth century, a brilliant emerald green. This colour was in great demand over the next two centuries and was used in everything from wallpaper to candy decorations and soap.

TOXIC EFFECTS:
As a poison, arsenic binds with sulphur-containing enzymes found in all cells of the human body; it can shut off cellular energy production, starve cells of protein and prevent them from repairing themselves. Arsenic delivered as a gas, called arsine, can destroy red blood cells. Arsenic appears in minuscule amounts in the human body, typically measured in millionths of a gram.

SYMPTOMS:
Chronic arsenic poisoning presents itself as itching, tenderness of the mouth, loss of appetite, nausea, diarrhoea and a swelling of the tissues – especially the face and eyelids – which results from an accumulation of fluids. The symptoms of acute arsenic poisoning are vomiting, diarrhoea, numbness in the feet, stomach pains, muscular cramps, intense thirst, urine suppression, collapse, coma and death.

TREATMENT:
The chelating agent dimercaprol can help the body naturally excrete arsenic, as it binds with the arsenic ions and helps the kidneys filter them out through the bloodstream.

Stomach pumping is also a common treatment, if administered quickly after the arsenic has been ingested.

FAMOUS POISONINGS:
- Frenchwoman Marie Lafarge famously poisoned her husband Charles with arsenic in 1840. She took the poison from the piles of arsenic used to kill rats around the house, and replaced it with white flour. Charles's body was exhumed twice to test for arsenic and was so badly decomposed that samples had to be taken with a spoon. Arsenic was discovered during Charles's second exhumation, and Marie was sentenced to life with hard labour.
- Charles Darwin took a regular daily dose of Fowler's Solution to treat tremors; this may have contributed to a chronic illness that bothered him for years but went undiagnosed.

OPPOSITE: A lump of arsenic in its pure metallic form as found in the Earth's crust.

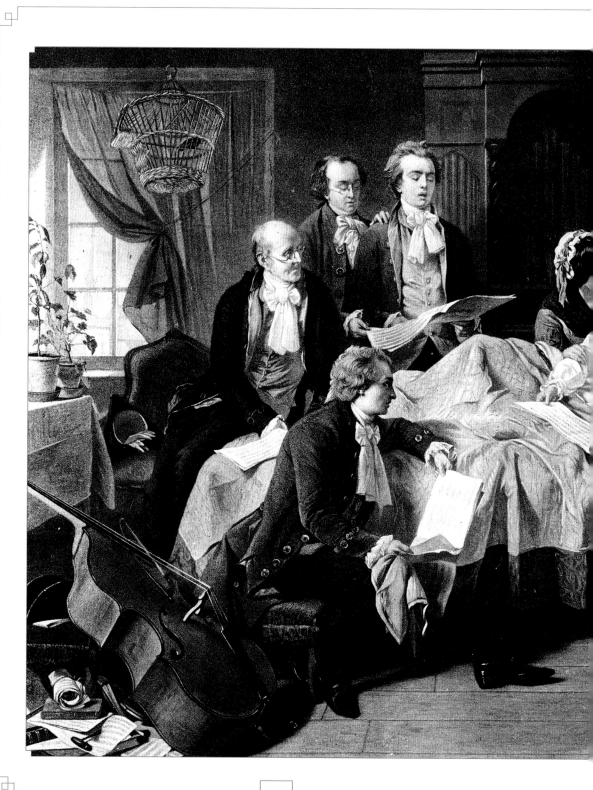

The Tofana Trap

As he lay on his deathbed in 1791, Wolfgang Amadeus Mozart became convinced that he was being poisoned.

"I will not last much longer," he said. "Someone has given me *acqua tofana* and calculated the precise time of my death." *Acqua tofana* was the notorious poison manufactured by Italian serial killer Giulia Tofana. She was responsible for at least 600 murders, but died a folk hero among women.

Renaissance Italy was a highly patriarchal society where women had few rights and little recourse against abusive husbands. Physical, sexual and emotional violence went unpunished, divorce was not possible and for many, death was the only solution.

Poison was a great leveller – it allowed aspiring widows to kill their spouses without violence. Giulia made it her business to help them. She developed the poison that bears her name, *acqua tofana*, and which, if administered in the correct way, was said to enable the murderer to choose their victim's precise time of death.

This was one aspect of the genius of *acqua tofana*. Its symptoms mirrored those of an advancing terminal disease, but the victim had time to put his affairs in order – and repent his sins, a point not to be underestimated in Catholic Italy. *Acqua tofana* was also colourless, odourless and undetectable. Mozart became convinced he was being slowly assassinated for this reason, despite the fact that Giulia Tofana had by then been dead for many years.

Details about Giulia are scant, but it is believed that she learned her craft from working as an apothecary's assistant in Sicily, where she was born. Poisoning, however, was a family affair, and Giulia's daughters were adopted into the business when they came of age. Giulia's mother had been executed for murdering her husband in 1633. This made Giulia sensitive to the needs of downtrodden women, especially those of lowly status, although she received referrals from women of every social position. She was known as a friend to the wife in troubled circumstances.

Giulia grew a thriving business from her cosmetics shops in Naples and Rome; it was from here that she sold *acqua tofana* secretly. The sale of cosmetics was a clever front, as the poison came as a powder and could be kept on a lady's dressing table without arousing suspicion. As Giulia's business progressed, so did the nature of the disguise. She made the *acqua tofana* in liquid form and sold it in vials as a healing ointment. Giulia's business thrived for over 50 years, but finally she was undone by a customer who lost her nerve.

In 1651, one of Giulia's clients added a lethal dose of *acqua tofana* to her husband's soup and then had a crisis of conscience. When she whipped the soup away from him, he demanded to know why. Eventually, the wife caved in and the whole story came out.

Soon the papal authorities in Rome were after Giulia. However, she was a popular figure, with many friends. For a time, she hid in a church. But when a rumour was started that Giulia had poisoned the city's water supply, efforts to apprehend her were stepped up. She was eventually caught and brought before interrogators. Under torture, Giulia confessed that she had helped to poison over 600 men between 1633 and 1651. It is unclear how accurate the figure is, as the poisoning could have been more widespread, or exaggerated under duress. Nevertheless, the verdict was inevitable. Giulia Tofana was executed along with her daughters and three female assistants in 1659. Her body was then thrown over the walls of the church that had given her sanctuary. Some of Giulia's accomplices were caught and then bricked up in the Palazzo Pucci's dungeons and left to die.

It is a testament to the legacy of *acqua tofana* that Mozart believed he was a victim of the poison more than 100 years after Giulia's death. No autopsy was ever performed on Mozart, and the cause of death was recorded as "severe miliary fever", a general term for fatal diseases. Medical experts today believe that Mozart may have had streptococcus, a bacterial infection.

ACQUA TOFANA

Acqua tofana was once described as a "harmless-looking liquid, a scant four to six drops of which are sufficient to destroy a man". Early symptoms were so slight that only a close relative might notice. The symptoms are described in the 1890 *Chambers's Journal*:

"Administered in wine or tea or some other liquid by the flattering traitress, [it] produced but a scarcely noticeable effect; the husband became a little out of sorts, felt weak and languid, so little indisposed that he would scarcely call in a medical man … After the second dose … this weakness and languor became more pronounced … The beautiful Medea who expressed so much anxiety for her husband's indisposition would scarcely be an object of suspicion, and perhaps would prepare her husband's food, as prescribed by the doctor, with her own fair hands. In this way the third drop would be administered, and would prostrate even the most vigorous man. The doctor would be completely puzzled to see that the apparently simple ailment did not surrender to his drugs, and while he would be still in the dark as to its nature, other doses would be given, until at length death would claim the victim for its own."

There was little that Renaissance medicine could do to save a victim of the poison, as they would be extremely unwell with vomiting, diarrhoea, an unquenchable thirst and a burning pain in the gut. Most of these symptoms are consistent with arsenic poisoning, although it has been suggested that *acqua tofana* may also have included belladonna, antimony, lead and cantharidin. It has been stated that the poison was not detectable in the blood, although autopsies and toxicological reports were not yet in use. The existence of *acqua tofana* was, of course, not known to authorities until Giulia Tofana was apprehended and the poison exposed.

ABOVE: Giulia Tofana mixes a toxic concoction in Evelyn De Morgan's 1903 The Love Potion.

The Catherine de' Medici Influence

When François I invited Catherine de' Medici to marry his son, he felt it would introduce some much-needed free-thinking Italian influence into his stuffy French court.

François was a king famous for moral squalor; no expense was spared for his debaucheries. Catherine was an apostle of this new morality: she was said to dabble in the dark arts and became known as "the king's whore".

From the moment he took the throne in 1515, François had railed against Catholic conservatism. He liked the idea of Martin Luther's new Protestantism and of living without boundaries. His preferred model was the noble houses of Renaissance Italy, where an opulent, decadent lifestyle was considered normal.

Italy was also famous for intrigue, scandal and murder. Catherine de' Medici, born in 1519 to a powerful Florentine family, was said to have single-handedly introduced Italy's poisoning tradition to France. Catherine tested poisons on the poor and sick, and during her lifetime was accused of poisoning her husband's brother and the Cardinal of Lorraine.

Catherine arrived at the French court as a flamboyant and unusual creature who shocked French sensibilities. She wore high heels normally favoured by prostitutes and surrounded herself with an entourage that included astrologers, alchemists and nine dwarves; the latter even travelled in their own miniature coach.

Catherine used sex as a way of securing deals and loyalties. Attending Catherine were around 80 ladies-in-waiting, known as her "flying squadron",

OPPOSITE: *Catherine de' Medici.*

who traded sexual favours for political gain. Catherine once threw a dinner party at which her flying squadron served the food topless before fulfilling requests for sex after the food was finished. Catherine also used poison to do away with anyone that displeased her. She killed one of her own astrologers in this way and then remarked that "he should have seen it coming".

If François enjoyed Catherine's intrigue, she was of no interest to his son Henry, her betrothed. Henry instead was infatuated with his mistress, whom he would fondle openly in front of Catherine. Part of the problem was Catherine's inability to produce children, a failing that she tried to correct with remedies such as drinking mule's urine or smearing cow's dung on her genitals. Nothing worked for 10 years, until, amazingly, Catherine produced not one but 10 children for Henry. Three would go on to be kings of France.

Henry, however, did not live to see this: he died at a jousting competition when a splintered lance put out his eye, leading to fatal septicaemia. Although Henry was wearing his mistress's colours, Catherine was said to grieve terribly for him.

After Henry's death, his son was crowned King Francis II. But his mother, Catherine de' Medici, was the real power behind the throne. Catherine oversaw one of the most turbulent periods in French history: the Wars of Religion. This conflict was fought between Protestant Huguenots and Catholics, and would divide the country in violence for more than three decades.

ABOVE: *The Venetian Council of Ten was formed in 1310.*

Catherine's policy of religious tolerance allowed Protestants the right to worship at home. She then brokered a marriage between Margaret, her Catholic daughter, and Henry, a Huguenot royal from the kingdom of Navarre. Catherine invited Henry's mother Jeanne d'Albert to the court and promised not to harm her children. D'Albert wrote in reply: "Pardon me if, reading that, I want to laugh, because you want to relieve me of a fear that I've never had. I've never thought that, as they say, you eat little children."

After visiting the court, d'Albert agreed to the marriage, as long as Henry was allowed to remain a Huguenot. This Catherine agreed to, but there was little love lost between the mothers. D'Albert reported of her meetings with Catherine: "I have spoken to the Queen three or four times. She only mocks me, and reports the contrary of what I have said to her, in order that my friends will blame me."

This is where the legend of the poisoned gloves

COUNCIL OF TEN

The great poisoners of Italy were Venice's Council of Ten, a tribunal tasked with averting plots and crimes against the Venetian state. To carry out their duty they ordered many assassinations with poison as the murder weapon. A killer was typically hired from another city, paid through an intermediary and funded by the council. Their dastardly deeds were recorded in a thin volume marked *Secretissima* (Top Top Secret). The book also included a proposed scheme whereby a Venetian doctor would create a poison from infected bubonic glands and spread it onto the insides of woollen caps to kill the Turkish enemy in Dalmatia.

FRENCH SCHOOL OF POISONERS

Catherine de' Medici is often accused of starting the poisoning craze in France, which by the mid-sixteenth century had become so entrenched that anyone of standing who died was assumed to have been poisoned. Some figures say the "French school of poisoners" may have created over 30,000 poisoners by the 1570s. This poison heyday was made possible because it was hard to differentiate poisoning from death due to disease, and there was no way of detecting poison in a victim's body.

enters the story. Among her many additions to French life, Catherine also introduced perfumed gloves. This was a highly popular item, as leather gloves had previously carried the odour of the urine and faeces with which they were treated. Italian gloves had, by contrast, been scented with herbs, spices and the oil of flowers such as jasmine, iris and orange blossom.

According to the story, Catherine sent d'Albert a pair of her gloves as a gesture of friendship before the wedding of their children. Shortly after wearing them, however, d'Albert died. Protestants quickly accused Catherine of killing their champion – a rumour that persists to this day. Catherine, they said, had added poison to the inside of the gloves.

Whether this tale has any merit or not, the wedding between Margaret and Henry went ahead. The day, however, is not remembered for the wedding but for a violent massacre of the attending Huguenots, known as the Saint Bartholomew's Day Massacre. It is thought that Catherine herself started the massacre by ordering the execution of Admiral Gaspard de Coligny, a possible contender to lead a Protestant uprising. The massacre was an orgy of savagery that resulted in the deaths of thousands. Catherine would survive the massacre, but not the opinions of the French public: she would be known as the corrupter of French sensibilities right up to her death. When this came, in 1589, Parisians wanted her body thrown into the Seine – the gravest of insults.

ABOVE: The 1572 Saint Bartholomew's Day Massacre.

Executing Elizabeth I

The court of Elizabeth I was a dangerous world of rebellions, conspiracies and assassination attempts. Many people wanted to kill the queen.

Restoring the Catholic Church, installing Elizabeth's sister Mary, or helping a foreign power invade England were popular plots. Poison was a common favourite for assassins of the day. And one such assassin was alleged to be Elizabeth's own physician.

Elizabeth I became used to attempts on her life, even from English nobles. Her sister Mary, Thomas Howard of Norfolk, Sir Francis Throckmorton and Sir Anthony Babington all conspired to kill Elizabeth. Pope Pius V also issued a Papal Bull in 1570 branding Elizabeth a heretic and threatening to excommunicate anyone who obeyed her. Elizabeth, then, was fair game for assassins.

In fact, there were so many plots against Elizabeth that a group of her loyal devotees led by Sir William Cecil and Sir Francis Walsingham signed an oath that they would put to death any who conspired against her. Walsingham, therefore, had the queen's best interest at heart when he recommended that she employ his personal physician, Roderigo Lopez.

Born in Portugal, Lopez had had particular success at London's St Bartholomew's Hospital and was well respected for his work in "dieting, purging and bleeding" – popular treatments of the day. Although achieving the position of Queen's physician, Lopez was considered untrustworthy among some at the royal court, especially as he was a Jew who had converted to Christianity. Before long, there were accusations against him.

The most serious of these was a "most dangerous and desperate treason" alleged by the Earl of Essex, Robert Devereux, a favourite of Elizabeth's, who disliked Lopez for mentioning his venereal disease to a mutual acquaintance. Essex's evidence was an intercepted letter in which Lopez was placing an order for pearls. One line in the letter that "the bearer will tell you the price of your pearls" was alleged to mean that Lopez's offer to assassinate the queen had been accepted and he should proceed forthwith.

Because of this accusation, Lopez was interrogated under torture and his house ransacked by spies. They admitted to Elizabeth: "In the poor man's house were found no kind of writings of intelligences whereof he is accused". Meanwhile, on the rack, Lopez confessed to accepting 50,000 crowns from the Spanish King to poison Elizabeth. He later retracted the statement, but it was too late: he was sentenced to be hung, drawn and quartered for the crime.

Even on the scaffold, Lopez protested his innocence and shouted that he loved Elizabeth as much as Jesus Christ, a comment that prompted a bout of anti-Semitic sniggering among the crowd. In Elizabethan England, personal freedom could be curtailed for national security and Lopez, a Jew and a foreigner, seemed a likely suspect. In the absence of hard evidence, suspicion and prejudice were enough to convict him.

ABOVE: A 1627 engraving of Roderigo Lopez with a Spanish conspirator.

ABOVE: Anthony Babington and six other conspirators are shown being hanged, drawn and quartered in 1586.

ABOVE: Opium poppy and seedpod.

Opium

*Opium is a narcotic drug extracted from the sap of seed pods from the opium poppy (*Papaver somniferum*). The poppy is grown in Turkey, Afghanistan, Burma, Colombia, Laos, Mexico, India, Pakistan and Thailand.*

OVERVIEW:
Raw opium has been extracted from the poppy seed in the same way since antiquity, by making an incision and letting the resulting sap turn yellow before scraping it off and drying it. The opium is then ground into powder, sold as lumps or treated to obtain drugs such as morphine, codeine and heroin, which make up the opiate family. Opiates are effective painkillers and also highly addictive; they are one of the most widely used and abused narcotics in history. The Sumerians called the poppy "the joy plant" in around 3400 BC; the ancient Greek physician Galen treated his patients with opium in around 150 AD, and the Romans introduced the drug to the pre-Arab East. From the sixth century AD, Arab traders introduced opium to China and India. In the sixteenth century, the physician Paracelsus introduced laudanum, a tincture of opium, back to the West. Opium was used as a painkiller and recreational narcotic until the mid- to late-twentieth century, when many Western countries banned its use except when prescribed by doctors.

TOXIC EFFECTS:
Raw opium can be drunk, swallowed, or smoked. In high doses, opium can cause respiratory and central nervous system depression, which can then lead to coma and death. The deprivation of oxygen caused by respiratory depression can lead to an inability to walk, followed by permanent damage to the brain and spinal cord. The habitual use of opium causes mental and physical deterioration in users and shortens the lifespan.

SYMPTOMS:
Opium typically affects the part of the brain that regulates breathing; breathlessness is often the first symptom of an overdose. Other symptoms include pinpoint pupils, nausea, vomiting, constipation, weak pulse, low blood pressure and dehydration, followed by cardiovascular depression, unresponsiveness, coma and death.

TREATMENT:
For opium overdoses, stomach pumping is followed by the ingestion of activated charcoal, which binds with poisons in the intestines before expelling them. The drug naloxone is a common antidote for opiate overdoses, as it reverses the depression of the respiratory and central nervous systems.

FAMOUS POISONINGS:
- Charles Dickens was a famous opium user who used the drug heavily until the time of his death in 1870, which was caused by a massive stroke. Opium in the nineteenth century was widely available as a painkiller and sold by tobacconists.
- Harold Shipman used the opium derivative morphine to poison as many as 250 victims in the late-twentieth century, making him one of the most prolific serial killers in history (pages 140–141).

BELOW: Bottles of nineteenth-century laudanum.

Chapter 3: Seventeenth- and Eighteenth-Century Poisons

By the seventeenth century, there were well-established schools of poison in both Italy and France. No one from any social class was exempt from the risk of being poisoned, but nobles and royals were especially involved.

In Paris, the wealthy libertine the Marquis de Sade proved Paracelsus' theory about dosages to be true when he poisoned four prostitutes with the alleged aphrodisiac cantharidin.

Paris was also home to "La Voisin", a purveyor of poisons whose customers included members of King Louis XIV's inner circle. Known as "inheritance powders", these poisons were a toxic brew of arsenic, belladonna and aconite; the rather less effective ingredients of bat's blood, sperm and the intestines of an infant were also sometimes added. Black magic rituals were used alongside inheritance powders in other attempts to assassinate the king.

The idea that poisoning was a supernatural act, which could be commissioned from demonic powers, was still a firmly held belief in the seventeenth century. Witchcraft was particularly feared. Central to the idea of European witchcraft was the use of poisonous plants including henbane, belladonna, datura and mandrake, used as hallucinogens in rituals and ceremonies. Often, these plants were made into ointments and rubbed onto sensitive areas such as the mucous membranes of the vagina and anus, which were first rubbed raw. The ointment was then applied via a "vehicle to be ridden", such as a broom handle, which was thought to give the witch the power to "fly to the witches' Sabbath" (page 10). The poison henbane, in particular, is known to put a person into a trance when taken as an intoxicant, and gives them the sensation of flying over the ground.

In Europe, tens of thousands of women were accused of witchcraft and burned at the stake for these crimes. To combat the witch-burning craze, Louis XIV issued an edict disowning deadly superstitions such as the belief that owning a black cat was proof that a person practised witchcraft.

As the fashion for executing witches dimmed in Europe, it reached fever pitch in the small American town of Salem, Massachusetts. Here a witch-hunt began among the town's own citizens when some of its daughters began exhibiting signs of the devil: strange babbling, fits and the unnatural contorting of the body. Accidental ergot poisoning may have been to blame.

Accidental poisoning is an enduring hazard; it struck down British soldiers in Jamestown, Virginia, after they fed on a salad of datura leaves, much to the mirth of the local colonists. Links between poisoning in the old world and the new would come to the fore in the following century. Murders on both sides of the Atlantic revealed the re-emergence of an old favourite: arsenic.

OPPOSITE: Here, in a black magic ritual, witches prepare potions and fly through the air.

Dosing Marquis de Sade

One June evening in 1772, the Marquis de Sade set out to procure several young Marseilles prostitutes for a night of sex and violence.

To make the women more amenable to the sodomy, whipping and debasement that followed, de Sade offered a crystal box filled with cantharidin-coated chocolates. But rather than act as an aphrodisiac as planned, the drug poisoned the prostitutes.

To escape the police, de Sade and Latour, his valet and partner in the crime of sodomy, fled Marseilles. It was not de Sade's first sex crime: the French noble's depravity had been exposed during the Rose Keller scandal four years earlier.

Keller was a 36-year-old baker's widow reduced to begging in Paris's Place des Victoires, where de Sade noticed her while standing under a statue of Louis XIV. Dressed in a grey jacket with white cuffs and holding a cane, de Sade approached the woman and offered to pay her for doing domestic work. He then drove her to his house in Arcueil and held a knife to her throat, forcing her to undress and tying her face down on his bed. He whipped her, poured hot wax into her wounds and sexually violated her. When de Sade untied her to put ointment on her wounds, Keller escaped through the second-storey bedroom and ran to a nearby village. Here, she lifted her skirts to reveal her injuries to some astonished village women. Authorities then issued a warrant for de Sade's arrest.

De Sade, however, avoided a long prison sentence after his powerful family agreed to contain de Sade's conduct and pay Keller off. Such was the sway held by the aristocracy in pre-Revolutionary France; money and influence could buy their way out of far worse crimes than the kidnapping and torture of a beggar. And de Sade's behaviour was perhaps of little surprise to his family.

Born Donatien Alphonse François, Marquis de Sade on 2 June 1740, the child de Sade became obsessed with whipping after being punished in this way at his Jesuit boarding school. His habit of flagellation and violence during sex, vividly described in his writing, gave rise to the term 'sadism': a psychosexual disorder where sexual arousal is experienced through the infliction of pain on others.

After school and a nine-year military career, de Sade married Renée-Pélagie Montreuil, daughter of a wealthy magistrate. He became famous as a libertine specializing in sexual cruelty with prostitutes, whom he began frequenting only weeks after his honeymoon. On the night of the Marseilles episode, de Sade had ordered Latour to find four or five "very young" prostitutes. The cantharidin candies he offered them would remove all inhibitions, or so de Sade thought. Instead, two of them suffered severe abdominal pains, a pressing desire to urinate and the vomiting

ABOVE: De Sade was a magistrate's son and libertine.

OPPOSITE: The Marquis de Sade is pictured surrounded by horned, winged demons.

of a black liquid. The next day, two more prostitutes told police that a nobleman had tried to poison them with chocolates from a crystal box.

De Sade and Latour escaped to Italy as a French court charged them with sodomy, attempted poisoning and moral outrage. They were found guilty and two straw effigies strung from a gallows in their absence. The men were later caught and jailed, but soon escaped and holed up in de Sade's chateau at Lacoste. The sadism didn't stop. De Sade carried out a six-week reign of terror against half a dozen teenagers trapped in his dungeon, much of the torture carried out under the indulgent gaze of his wife, Renée-Pélagie.

The local villagers knew de Sade as a "werewolf"; one tried to shoot the Marquis after an incident involving his daughter. De Sade's exploits were finally too much for his mother-in-law, Madame Montreuil, who had the king sign a *lettre de cachet*, a warrant to keep de Sade permanently incarcerated. De Sade was arrested in 1777 and would spend 29 years in prisons and an insane asylum.

While imprisoned in Paris's Château de Vincennes and the Bastille, de Sade devoted himself to writing erotic literature which he himself called unpublishable because of its extremity. Although de Sade was allowed an aristocratic incarceration, with books, food and other comforts, he was freed during the 1789 storming of the Bastille and given official positions under the revolutionary National Convention. After a period of freedom de Sade was arrested in 1801 and imprisoned without trial for the anonymous publication of his novels *Justine* and *Juliette*, condemned as "abominable" by Emperor Napoleon Bonaparte.

De Sade ended his life in the Charenton insane asylum, where he wrote and staged his plays with fellow inmates. He died at 74, bankrupt and unrepentant; his last wish not to be given a Christian burial was ignored. His body was later exhumed so a phrenological study could be made of his skull, a plaster copy of which is now in Paris's anthropological Museum of Man, alongside those of revolutionary leader Maximilien Robespierre and Napoleon.

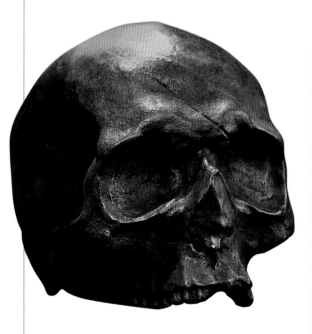

ABOVE: *A bronze cast of the skull of the Marquis de Sade.*

POISON OR APHRODISIAC?

Cantharides, or Spanish Fly, is one of many substances used as sexual stimulants since antiquity. Aphrodisiac herbs mentioned on ancient Assyrian cuneiform tablets include stinging nettle, red poppy and asafoetida, a pungent resin used as a cooking spice. Ancient Indian Hindu aphrodisiacs include crocodile eggs, burnt pearls and lizards' eyes. In China, powdered rhinoceros horn is still used. In Africa, the traditional aphrodisiac Yohimbine, an extract from the Yohimbé tree, stimulates the nervous system and increases erectile function. However, high doses of Yohimbine cause sweating, nausea, vomiting and death. This is also the case with Cantharides. Like many unregulated and misunderstood substances, too high a dose can make a poison of a lover's medicine.

ABOVE: Letter from the Marquis de Sade to Monseigneur Fouché, Chief of Police, requesting freedom to leave his prison cell to visit Paris.

Cantharides

Cantharides is extracted from blister beetles, particularly the emerald-green blister beetle (*Lytta vesicatoria*), also known as the Spanish Fly. This gives the poison its common name.

OVERVIEW:

Used since antiquity as an aphrodisiac, cantharides is a powder created from the beetles' dried, crushed carcasses. The poison's principal ingredient is cantharidin, a chemical secretion which protects the beetles against predators. In humans, cantharidin causes the widening of blood vessels and the irritation of the urinary bladder and urethra. Cantharides creates strong and powerful erections in men and, in women, a genital itching that is claimed to be eased by intercourse. History has shown that cantharides is dangerous and unreliable: the Marquis de Sade poisoned two prostitutes with it at an orgy, and high doses of cantharides have caused many deaths. A side effect of cantharides is an abnormally prolonged erection which can lead to permanent penile damage and erectile dysfunction. Today, cantharidin is used as a topical medicine in small doses and cantharides listed as a 'problem drug' only administered under strict medical supervision.

TOXIC EFFECTS:

Cantharides is a vesicant, or blistering agent, that can be fatal in doses of 10ml or over. The same blistering that appears on human skin after brief contact with the poison also occurs internally if swallowed. This can lead to extreme haemorrhaging in the intestinal tract and renal dysfunction. When the chemical cantharidin was first isolated in 1810, chemist Pierre-Jean Robiquet announced it was a violent poison second only to strychnine. A test for cantharides as a poison was developed in 1880. It involved reducing an aqueous stomach sample from the deceased over heat. This condensed extract was then placed on healthy skin and bandaged. If the skin blistered after a few hours it indicated the presence of cantharides.

SYMPTOMS:

Cantharidin causes irritation to any organic matter it touches. If swallowed, even 2ml of the toxin can cause burning in the mouth, throat and stomach, followed by vomiting and bloody diarrhoea. Higher doses can lead to heart and kidney failure, coma and death.

TREATMENT:

There is no antidote for cantharides. Drinking water can help dilute the poison.

FAMOUS POISONINGS:

- London office manager Arthur Ford accidentally murdered workmate Betty Grant in 1954 by lacing her sweets with cantharides in an attempt to seduce her. Their colleague, Miss Malins, also ate a sweet and died. Ford confessed and was jailed for five years.
- According to the ancient Roman writer Tacitus, Augustus' wife Livia laced her dinner guests' food with cantharides so they would commit public sexual indiscretions and could be later blackmailed.

ABOVE: *"Aphrodisiac" extracted from the cantharides beetle.*

OPPOSITE: *The emerald-green blister fly.*

Affair of the Poisons

In 1676, Madame de Brinvilliers of Louis XIV's royal court was found guilty of murder through poisoning.

Under the water cure, a torture that stretched a victim's body over a wheel before filling it with water, de Brinvilliers confessed to killing her husband, father and brothers to inherit their estates. Before she was beheaded and burned at the stake, de Brinvilliers implicated other nobles at court. The Affair of the Poisons had begun.

Madame de Brinvilliers' confession sent shockwaves through Louis XIV's court, which had been struck by several mysterious deaths. Now suspecting foul play, the Sun King immediately established a commission, the *Chambre Ardente* (burning chamber) to investigate. Heading the commission was chief of the Paris police Gabriel Nicolas de la Reynie, who quickly extracted confessions from fortune-tellers, alchemists and ex-priests about poisoners selling aphrodisiacs to a number of French nobles.

Most prominent among the names given was Catherine Monvoisin, nicknamed "La Voisin", a self-styled purveyor of poisons and alleged sorceress. After arrest and interrogation, La Voisin admitted to the wholesale distribution of poisons, potions and powders to nobles including those in Louis's inner circle. Helpfully, La Voisin had kept a list of her clients, which, disturbingly for Louis, revealed among them his former mistress Madame de Montespan.

The king had always kept mistresses, but all had been swept aside by Montespan, an enchantress of beauty and brains who had won the heart of the king. Louis had lavished jewels, apartments and affection upon Montespan to the exclusion of all others – that is, until she gave birth to the final of her seven bastard offspring with the king. Now a mother of nine, Montespan had been unable to regain her prenatal figure.

It was only a matter of time before Louis tired of his corpulent mistress and sought a new favourite; it was no surprise that his gaze fell upon Mademoiselle de Fontanges. Described as being "as beautiful as an angel and as stupid as a basket", de Fontanges made a great impression upon the king. He began wearing jackets trimmed with ribbons that matched her dresses, endowed a great annual pension upon her, and gave her a new title. Before long, women of the royal court were wearing their hair "Fontanges" style, and the new mistress herself was pregnant. De Montespan described the Fontanges hairstyle as "in bad taste" and smouldered with rage and resentment. This was the moment, according to la Voisin's testimony, that de Montespan decided to poison not only de Fontanges but Louis as well.

In truth, it was not the first time de Montespan had resorted to mind-altering substances to better her position: she had been a regular client of La Voisin's for some time. La Voisin told the *Chambre Ardente* that de Montespan had been buying aphrodisiacs from her for many years and even performed black magic rituals to secure the king's favour. Then, after the news of de Fontanges' pregnancy, de Montespan had offered to pay La Voisin a small fortune to murder both the mistress and the king.

The method La Voisin devised for the murder of de Fontanges was to present her with a poisoned dress and a pair of poisoned gloves. For the king, poison would be soaked into a petition, a typical letter of request to the king from a subject. But before La Voisin could bring the plot to fruition, she was arrested by Gabriel Nicolas de la Reynie in connection with the Brinvilliers' poisonings.

OPPOSITE: *Madame de Brinvilliers' water cure.*

ABOVE: Catherine Monvoisin, aka 'la Voisin'.

De Montespan's murderous plans were then quickly exposed.

La Voisin's confessions and those of her daughter led to a frenzied witch-hunt by the *Chambre Ardente* and the arrest and execution of 36 people believed to be involved with magic or poison. La Voisin herself was burned at the stake. About 34 high-ranking aristocrats were sent into exile, and Louis issued a royal edict in 1682 banning magic, poisons and any person who used them.

As for de Montespan, news that she had fed love potions to Louis made sense to the king. He had often suffered from physical ailments such as headaches after an evening with de Montespan. But despite the evidence, he could not allow his former mistress and mother of several of his children to be burned at the stake. She was instead exiled to a convent for the remaining 17 years of her life.

Mademoiselle de Fontanges did not bear Louis any children, but instead suffered two miscarriages and then died after a mysterious illness that lasted for over a year. The royal court naturally assumed de Montespan's assassination plan had finally borne fruit, but the autopsy showed no signs of poison. However, no amount of evidence would convince commentators of the time, nor many of those since, who have suspected de Fontanges was murdered by poison.

ABOVE: *Louis XIV and Madame de Montespan.*

POISONERS AND PURCHASERS

The Affair of the Poisons exposed a criminal underworld that dealt primarily in the manufacture and sale of love potions, magic charms, spells and poisons commonly known as "inheritance powders". These were used by prominent nobles, including the Duke of Buckingham and the Marshal of Luxembourg, for seduction, financial gain and the elimination of rivals. At the centre of these activities was the sorceress, fortune-teller and abortionist La Voisin, who dressed in an ornate robe of red velvet and embroidered gold eagles and was assisted by a deformed ex-priest called Étienne Guibourg. Although the order was given to torture a confession from La Voisin, she was instead fed alcohol during her interrogation, which was enough to loosen her tongue. After her execution, La Voisin's daughter further aided the *Chambre Ardente* in its investigation.

INHERITANCE POWDERS

The so-called "inheritance powders" provided to the nobles of Louis XIV's court were believed to be a mixture of arsenic, aconite, belladonna and opium. This concoction would have been loosely based on *acqua tofana* (pages 55–56), which was created in Italy only a few decades earlier. The love potions administered to Louis XIV by de Montespan included bat blood, cantharidin, iron filings, menstrual blood and sperm. One mixture was said to have consisted of a baby's intestines, blood and bones, ground up after a black magic ritual to sacrifice the infant. Étienne Guibourg confirmed that he had been present at the ritual with de Montespan and La Voisin; a crucifix was held on de Montespan's naked stomach as the infant was disembowelled.

Salem Witch Trials

In January 1692, three young girls of Salem Village, Massachusetts, began displaying strange symptoms thought to show that they were possessed by the devil.

Elizabeth Parris, Abigail Williams and Ann Putnam would fall into fits, scream, speak in tongues and contort their bodies into unnatural positions. When interrogated, the girls blamed a Caribbean slave, a homeless beggar and an elderly bedridden woman for the symptoms. The verdict: witchcraft.

Until the seventeenth century, witchcraft hunts had been mainly a European phenomenon. Between the fourteenth and eighteenth centuries, tens of thousands of women were executed in Europe for this crime. However, just as the European witchcraft craze dissipated, it was picked up by the small and troubled community of Salem Village.

Salem was a town divided between rich and poor and was struggling to cope with an influx of refugees from King William III of England's war with France. A further schism emerged between those who supported a controversial new minister, Samuel Parris, and those who did not.

It was Parris who forced a confession out of his daughter Elizabeth and niece Abigail that the family slave Tituba had bewitched them, along with the beggar Sarah Good and widow Sarah Osborn. Osborn and Good denied the charges, but Tituba, after repeated badgering, told her interrogators what she thought they wanted to hear: that she had made a pact with the devil who made her sign his book. This is where the names Good, Osborn and seven others had been revealed to her.

Hysteria ensued. Several more Salem girls began experiencing fits; dozens of others were accused of witchcraft, some of them prominent townswomen. Opinion divided along the town's traditional lines. The accused were forced to defend themselves without legal counsel before seven judges. The witchcraft trials would last several weeks.

The admission of "spectral evidence" meant that the victims could point out the accused, whose "spectres" had allegedly pinched and bitten them. To add further drama, the victims whimpered, babbled and writhed every time an accused took the stand. Those who confessed or named other witches were spared execution; those who insisted on their innocence received harsh punishments. Few dared to speak out against the injustices of the trial. Nineteen women were hanged, and an elderly man was pressed to death beneath heavy stones. Five more died in custody.

In 1702 the authorities declared the Salem witch trials unlawful. During the soul-searching that followed, some sought a scientific diagnosis for the conditions affecting the Salem girls. One later theory was that they were suffering from ergot poisoning, caused by eating rye infected with the fungus ergot. This has been known to cause fits, choking and hallucinations. It cannot, however, excuse the actions of Salem Village, whose leaders convicted and killed their own citizens without lawful trial.

OPPOSITE: *A fanciful nineteenth-century rendering of the Salem witch trials.*

Ergot

Ergot is a fungal mould of the genus Claviceps *that infects wheat, rye, barley and other grains. Its most common member is rye ergot (*Claviceps purpurea*), which causes ergot poisoning in humans.*

OVERVIEW:

Today, our view of medieval Europe is of a continent riddled with disease. The Black Death, St Anthony's Fire and St Vitus's Dance periodically tore through local towns and villages, decimating their populations. Victims of St Anthony's Fire were struck by a violent madness and a blackening of limbs that led to the loss of fingers, toes, hands and feet. St Vitus's Dance caused convulsions, hallucinations, sensations of suffocation and extreme twitching that gave them the appearance of being possessed by demonic spirits. We now know that these two diseases, and others of the Middle Ages – perhaps even the Black Death – were actually caused by ergotism, poisoning from rye ergot. It has also been established that places in northern Europe and America where witchcraft was widely reported also had rye as the staple cereal. It seems likely that instead of suffering from witchery or the influence of the devil, those persecuted in the witch-hunts were actually victims of ergotism.

TOXIC EFFECTS:

The two types of ergotism are convulsive ergotism and gangrenous ergotism. Both are caused by the four types of poisonous fungi alkaloids, called mycotoxins, that ergot produces. Any of the mycotoxins can trigger the hallucinations that are part of convulsive ergotism: these can cause visions of ferocious animals and waterfalls of blood, alongside extreme convulsions. Gangrenous ergotism cuts off the supply of blood via blood vessels, leading to the death of tissue, infection and gangrene, followed by the loss of limbs. Death is the last stage of extreme ergotism.

OPPOSITE: *A magnification of the ergot fungal mould.*

SYMPTOMS:

Symptoms of convulsive ergotism include spasm, seizures, diarrhoea, chronic itching, writhing, staggering, nausea and vomiting, followed by delusions and hallucinations. Symptoms of the more common gangrenous ergotism include numbness, muscle twitching, a weak pulse, burning sensations and blackening of the limbs, before a loss of sensation altogether and gangrene. It is unusual to suffer from both convulsive ergotism and gangrenous ergotism at the same time.

TREATMENT:

There is no antidote for ergotism. Treatment is given for the symptoms by administering tranquillizers for the convulsions and anti-coagulants and vasodilators to restore blood flow.

FAMOUS POISONINGS:

- In the Viking saga *Heimskringla*, it was recorded that Norwegian king Magnus Haraldsson died from ergotism a few years after the Battle of Hastings.
- In 1951, 4,000 villagers from Pont Saint Esprit in France were reported to have gone mad with hallucinations, vomiting, burning sensations in their limbs and convulsions. Ergotism was found to be the cause.
- A European outbreak of the plague that caused blisters and rotting of limbs in 857 is thought to have resulted from ergotism. In addition, some people have speculated that the Black Death, which decimated the population of Europe in the fourteenth century, was caused not by the bubonic plaque but by ergotism.

The Jamestown Poisonings

In 1676, King Charles II of England received troubling news from his American colonies. In Virginia there had been an armed uprising and its capital, Jamestown, had been razed.

To put down the rebels and restore order, Charles dispatched his troops. However, their effectiveness was hampered by ingestion of the poisonous plant *Datura stramonium*, commonly known afterwards as Jamestown Weed.

The trouble in Virginia began when governor Sir William Berkeley refused to order violent retaliation for native American raids on frontier settlements. An angry meeting was convened by Nathaniel Bacon, who supplied the attending settlers with brandy and was subsequently elected their leader. Bacon ordered the attack and massacre of a local Susquehannock village and then demanded that Berkeley give him a commission to lead militia against further native Americans in the area. When this request was denied, Bacon's Rebellion began. Some months later, around 500 rebels burned Jamestown to the ground.

Luckily for the unwitting British soldiers dispatched to subdue the settlers in 1676, Bacon's Rebellion did not last long. This was a fortunate twist of fate; less fortunate was that the soldiers had eaten large amounts of datura, believing the plant to be a local vegetable. The effects of the poison were recounted by Robert Beverley in his 1705 book *The History and Present State of Virginia*:

"Some of the soldiers sent thither to quell the rebellion of Bacon … ate plentifully of it [datura], the effect of which was a very pleasant comedy, for they turned natural fools upon it for several days: one would blow up a feather in the air; another would dart straws at it with much fury; and another, stark naked, was sitting up in a corner like a monkey, grinning and making mows [faces] at them; a fourth would fondly kiss and paw his companions, and sneer in their faces with a countenance more antic than any in a Dutch droll.

"In this frantic condition they were confined, lest they should, in their folly, destroy themselves – though it was observed that all their actions were full of innocence and good nature. Indeed, they were not very cleanly; for they would have wallowed in their own excrements, if they had not been prevented. A thousand such simple tricks they played, and after eleven days returned themselves again, not remembering anything that had passed."

The soldiers recovered from their stupor without any lasting effects. As for Bacon's Rebellion, its eponymous leader died of dysentery that same year and the rebel settlers returned to farming a few months later. However, many considered the insurrection a precursor to the American Revolution and subsequent independence from Britain.

OPPOSITE: Jamestown, Virginia burns during Bacon's Rebellion of 1676.

Datura

Datura stramonium is one of nine species of poisonous flowering plants of the Solanaceae family. It is commonly known as Jamestown weed, jimsonweed, thorn apple, devil's trumpet and stink weed.

OVERVIEW:

Arguably responsible for more poisonings than any other plant, datura is a hardy, pale green herb that grows to around five feet tall and sprouts greenish-purple leaves and white and purple flowers. Its highly poisonous seeds are contained within hard spiny capsules sometimes called thorn apples. Although datura seeds and leaves have an unpleasant taste, they are often mixed with farming feed, leading to the accidental poisoning of farm animals. The datura plant was harvested as a tea or tobacco for its hallucinogenic properties by the indigenous peoples of the Americas and is commonly misused by curious teenagers today. The narcotic effects of the plant can last for over two days and at high doses are seldom described in favourable terms. Datura intoxication can feature hallucinations of talking objects, the unexpected appearance of non-existent people and the smoking of phantom cigarettes. The name datura may have originated in ancient India, where a criminal gang called the *dhatureas* used the plant to drug their victims. It has been suggested that datura was used to induce the visions of ancient Greece's Oracle at Delphi and was also fed to sacrificial victims of the Aztecs before their hearts were torn out.

TOXIC EFFECTS:

The three toxic chemicals in datura are hyoscyamine, atropine and scopolamine. These have powerful anticholinergic properties that block the actions of neurotransmitters between cells and therefore affect how the body functions. Anticholinergic poisoning can occur after 30 minutes of ingesting the poison and at high doses may quickly result in death. As little as half a teaspoon of datura seeds can lead to seizures and cardiac arrest, although the strength and toxicity of the seeds varies among species, seasons and locations, which makes a safe dosage notoriously hard to calculate.

OPPOSITE: Illustrations of the Datura stramonium plant.

SYMPTOMS:

Symptoms of datura poisoning have been described as "red as a beet, dry as a bone, blind as a bat, mad as a hatter". These include a dry mouth, excessive thirst, hot skin, headache, slurred speech, impaired coordination, vertigo, weak pulse, disorientation, urinary retention, constipation, hallucinations, convulsions, coma and death. Hospitalizations and fatalities from datura overdoses, either accidental or deliberate, are common.

TREATMENT:

The drug physostigmine is sometimes given when severe anticholinergic toxicity has followed datura poisoning. Charcoal is also used to decontaminate the gastrointestinal tract, and sedatives are administered.

FAMOUS POISONINGS:

- In AD 38, Roman general Mark Antony and his army were retreating from Parthia and could find little to eat. Some of his soldiers ate datura and as a result resorted to "turning over every stone in their path with the greatest gravity, as though it were a difficult task". The saying "leave no stone unturned" is thought to have originated from this incident.
- In 2008, a family from Maryland, USA, were found confused, laughing uncontrollably, vomiting and hallucinating. Thirty minutes later, two family members fell unconscious and the others became aggressive, incoherent and disorientated. They had eaten a stew accidentally seasoned with datura.

Murdering Mary Blandy

Mary Blandy did not seem like a poisoner capable of parricide. She was middle-class and well-educated, and had been brought up in the leafy English suburbs of Henley-on-Thames, Oxfordshire.

Mary's father, Francis, a successful lawyer, only wanted the best for his daughter. But his plans to find her a suitor would end in his own death by poisoning and Mary's execution by hanging.

In 1746, Mary Blandy was unmarried and in her late 20s. Her father feared that Mary would be left permanently on the shelf and hit on a novel idea of finding a husband. He advertised for suitors in the local newspaper and offered the handsome sum of £10,000 as a dowry. Many men came forward, but the only one thought worthy of Mary's affections was a certain William Henry Cranstoun.

As an army captain and the son of a Scottish noble, Cranstoun held a certain cachet. Unfortunately, his physical characteristics did not match his social position: he was short, had mean features and a face heavily scarred by smallpox, and was 12 years older than Mary. Cranstoun was also, it emerged after a period of engagement to Mary, already married. Francis became livid at this news and broke his daughter's engagement, despite assurances from Cranstoun that his previous marriage was not legal.

The wedding was cancelled, but Mary continued to see Cranstoun in secret. Francis, however, was adamant that he would not allow a bigamist into the family. How, then, to break the impasse? Cranstoun devised a plan that used a white powder, which he described as a love tonic. By surreptitiously feeding the powder to her father, Mary would supposedly make him forgive her Scottish suitor and make sure that the two were united in marriage.

However, Cranstoun's tonic was in fact arsenic, and by adding it to her father's porridge Mary was actually beginning to poison him. Francis died on 14 August 1751, and Cranstoun quickly disappeared. Unhappily for Mary, Francis' doctor, Anthony Addington, had the foresight to test some of the left-over powder by burning it. This revealed a smell of garlic, which was known to indicate the presence of arsenic. Addington's test, therefore, became one of the first forensic studies for the use of poison as a murder weapon.

Mary was hanged on 6 April 1752, after being variously described as a "confused, lovesick fool" who didn't knowingly murder her father, and a "cold-blooded killer", who did. Perhaps the greatest irony is that the marriage would not have concluded with a £10,000 dowry payment to Cranstoun; Francis Blandy died with a total of £4,000 in his bank account.

ABOVE: The preserved stomach of a patient with hemorrhagic gastritis, whose death was attributed to acute arsenic poisoning.

ABOVE: Mary Blandy's last words at her execution were allegedly: "for the sake of decency, gentlemen, don't hand me high."

Chapter 4: Nineteenth-Century Poisons

The Victorian era was the Golden Age of poisoning. Poisons were widely available, and the introduction of life insurance provided a profitable new motive for their use.

Suddenly a price had been put on a life, and for many it was worth the surreptitious murder of a spouse or family member. Poison was the perfect method as it was difficult to detect and could be used equally well by both wives and husbands. The favourite poison was arsenic.

Arsenic was the great, ubiquitous poison of the nineteenth century, and it was used in everything in England from rat poison to wallpaper. Anyone could obtain arsenic from the local shop, as freely as they could buy bread or milk. There was consequently no more popular way to murder someone than through a tonic of arsenic.

Murder by poisoning rose steadily from the 1820s and peaked around 1850. It is not surprising that the new methods of forensic toxicology came to be developed rapidly at the same time. In 1832, James Marsh, a chemist at London's Royal Arsenal, Woolwich, was asked to test for arsenic in the poisoned body of George Bodle. Unfortunately the stomach sample was too old, and George's accused grandson John went free. John later confessed to the crime. Marsh was so infuriated with this that he perfected an effective test for arsenic that subsequently became known as the Marsh Test.

The Marsh Test caught the attention of chemists worldwide; the German chemist Karl Friedrich said it was a game-changer. The press soon learned how alluring the game was. A flamboyant Spanish chemist, Mathieu Orfila, used the Marsh Test to establish the presence of arsenic in the body of Charles Lafarge in 1840. Charles's wife Marie then became the first person to be convicted as a result of forensic toxicological evidence. She was sentenced to life imprisonment, and the Marsh Test was widely touted as the great exposer of a poisoner's dark deeds.

From that point on, forensic toxicological evidence could tell who had died naturally, and who had been poisoned. Poisoning would never be the same again.

The story of poisoning thereafter became a search by the murderers for new poisons that could not be detected. In the twentieth century, this led the Soviet Union to develop unknown poisons such novichok, which for a time flew under all the toxicological radars. However, towards of the end of the Victorian era, many others continued to try their luck with arsenic as well as strychnine, and with a new poison that would be put to terrible use a few decades later: cyanide.

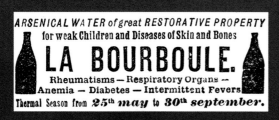

ABOVE: La Bourboule was a mixture of arsenic and French mineral water sold as a cure-all in the nineteenth century.

ABOVE: James Marsh was a British chemist who devised the first dependable test for detecting arsenic.

Marrying Florence Maybrick

The conviction of Florence Elizabeth Maybrick for murdering her husband with arsenic caused an outcry. There was a near riot on 7 August 1889, the day of the verdict:

A report from *The Illustrated Police News* described how "the crowd assembled outside the court, cheered the prisoner, hissed the jury, and hooted the judge – came very near lynching him". Popular opinion was with Florence; but was she guilty?

James Maybrick was a well-known hypochondriac who ingested various tonics of the day, including strychnine, cyanide, henbane and morphine. He also took a daily measure of Fowler's Solution, which contained arsenic, for his flagging libido. James was a philanderer who had five children by one of his mistresses after marrying Florence, an American teenager 23 years his junior, whom he met on a liner bound for Britain in 1880.

The marriage between Florence and James floundered after she found out about his affairs; in response, Florence took her own lover, which James subsequently discovered. From that point on the couple argued openly, and James even blackened her eye one day at the races in London. According to her testimony at the trial, Florence then purchased arsenic-based wallpaper, which she soaked in water to extract the poison. This was, she said, to use in the preparation of make-up. However, the prosecution alleged that she used the poison to kill James.

James fell ill in April 1889 after taking a tonic that contained strychnine. He died on 11 May. An examination of James's body ordered by his brothers found traces of arsenic. A police investigation subsequently revealed that a chemist had prescribed arsenic to James several times; enough of this tonic was found in the house to kill 50 people. Florence's adultery, however, counted against her, as the judge told the jury that this meant she was no better than a murderess anyway. Florence was sentenced to hang.

However, after a public outcry on both sides of the Atlantic, Florence's verdict was reviewed. The British Home Secretary and Chancellor both signed a letter stating that "The evidence clearly establishes that Mrs Maybrick administered poison to her husband with intent to murder; but that there is ground for reasonable doubt whether the arsenic so administered was in fact the cause of his death."

Florence's sentence was thereafter commuted to life imprisonment. She served nine months in solitary confinement and then with the general prison population under the "silent system", which forbade talking at all times. After her release in 1904, Florence moved back to the United States, where she protested her innocence on the lecture circuit. She died penniless and alone in a squalid Connecticut house in 1941.

LEFT: *Fowlers Solution was a nineteenth-century arsenic remedy used to treat psoriasis.*

ABOVE: *Florence Elizabeth Maybrick never stopped protesting her innocence. Her 1905 book,* My Fifteen Lost Years, *recounted her experiences.*

Bradford Sweets Poisoning

On Sunday morning, 30 October 1858, a strange affliction fell over the English city of Bradford. Two boys died suddenly, and several others were struck with a strange, violent illness.

Some blamed an outbreak of cholera; others, the plague. As the day wore on, more deaths were reported. Suspecting foul play, the police quickly discovered the contributing cause: peppermint humbugs.

A few days earlier, William Hardaker, known locally as Humbug Billy, set up his regular sweet stall at Bradford's Green Market. He was selling the popular peppermint humbugs as usual, although this batch had been bought at a discount as Hardaker hadn't liked the colour. Hardaker bought his sweets from wholesale confectionery dealer Joseph Neal, who happened to be in the habit of swapping sugar for a substance called "daft".

Daft was a mixture that typically contained plaster of Paris, powdered limestone, sulphate of lime, or any other harmless substance that could adulterate sweets in place of expensive sugar. Unfortunately, Joseph Neal's recent order for 12 pounds of daft had been fulfilled by a young assistant at the local chemist. Instead of weighing out 12 pounds of daft, the young man had weighed out 12 pounds of arsenic trioxide from the container next to it. Both were white powders of a similar consistency.

The sweet-maker working for Joseph Neal felt that there was something wrong with the subsequent batch of humbugs made with the chemist's daft. However, this didn't prevent him from eating one himself, or selling the batch to William Hardaker, who also consumed one. Both men fell ill soon afterwards. Because Hardaker had purchased the humbugs at a discount, he also sold them at a discount; his humbugs did a roaring trade that day.

At midnight on Sunday, the district bell-ringer woke Bradford with warning cries about the deadly humbugs. However, by that time over 200 people had fallen ill and 20 later died from arsenic poisoning. It emerged that each humbug had enough arsenic to kill two adults. The next day the chemist, his assistant and Joseph Neal all appeared before the court on charges of manslaughter; none, however, went to jail.

Despite this, the assistant's tragic mistake did lead to the 1860 Adulteration of Food and Drink Bill, which set out which ingredients could be combined to make sweets. The Bradford sweets poisoning also led to the UK Pharmacy Act of 1868, which introduced stringent regulations for chemists on the handling and selling of named poisons. Finally, the 1874 abolition of the sugar tax made sugar an affordable substance for all.

OPPOSITE: This cartoon about the Bradford Sweets poisoning appeared in Punch magazine in 1858. Food adulteration was a major problem in Victorian England.

THE GREAT LOZENGE-MAKER.

Mothering Mary Ann Cotton

It was the sudden death of Mary Ann Cotton's seven-year-old stepson that rang alarm bells. The boy had prevented Mary Ann from re-marrying, but she had casually mentioned to an official that "I won't be troubled long".

The subsequent investigation exposed Mary Ann as the deadliest serial poisoner of the Victorian era: she was the "black widow" who routinely murdered her husbands and children.

Mary Ann Robson was a miner's daughter born in county Durham in 1832. She worked as a nurse before marrying William Mowbray, husband No. 1, in 1852. Over the next decade, the couple had eight or nine children; all but three died of a common affliction known as gastric fever. "Gastric fever" was a generic term used to cover illnesses such as typhoid fever that commonly caused sudden infant death. Its symptoms also mirrored those of arsenic poisoning.

Infants, however, were not the only ones to die of gastric fever: William Mowbray succumbed to the disease in 1864, leaving Mary Ann with a life insurance payout of £35, or about six months of a labourer's salary. Mary Ann then left her remaining child with her mother and married husband No. 2, George Ward, in 1865. The marriage lasted a year before Ward died and Mary Ann received another life insurance payout.

Mary Ann then went to live with her mother, who died a week later, before returning to Durham to marry husband No. 3, James Robinson, in 1867. After the deaths of three more children, Robinson threw Mary Ann out when she asked him to take out a life insurance policy.

For a period Mary Ann was homeless, before committing bigamy by marrying husband No. 4, Frederick Cotton. During the marriage, Cotton, his sister, and three of his children – two with Mary Ann – all died. Mary Ann then took up with a former lover, Joseph Nattrass, while also falling pregnant to another man, John Quick-Manning. In 1872 Nattrass died, and Mary Ann told the official about not being "troubled long" by her stepson Charles Cotton so she could marry Quick-Manning.

After Charles died, the official alerted police and an autopsy found arsenic in the boy's stomach. Nattrass and some of Mary Ann's children were exhumed; all had died of arsenic poisoning. Investigators only charged Mary Ann for Charles's murder with arsenic, but it is believed that she had killed up to 21 people in this way. The press dubbed her the "black widow", and she was sentenced to hang in 1873.

As a grisly afternote to the Mary Ann story, the hangman's noose was made too short to break the woman's neck, and the executioner had to press on her shoulders to finish the job; it took a full three minutes for her to die.

OPPOSITE: Mary Ann Cotton is widely considered to be Britain's most prolific female serial killer.

The Matchstick Girls

In 1888 the English social reformer Annie Besant published a shocking article about a match factory in London's East End. Here, teenaged girls were being forced to work 14-hour days for meagre wages and routinely subjected to fines and violence from their foremen.

Worse still, the workers were constantly exposed to the deadly white phosphorus added to the match tips. This gave many of them a devastating disease known as "phossy jaw".

Besant's article carried the headline "White Slavery in London" and caused outrage not only among its readers but also from the owners of the match factory, Bryant and May. In Besant's article, conditions for the matchstick girls, many of them only 13, were laid bare.

The girls worked from 6.30 am until 6 pm, standing for the whole time, with only two breaks. They were not allowed to talk or sit while they worked and were only paid between four and nine shillings a week, depending on their age. From their earnings, two shillings were spent on rent and much of the rest on food, which for most consisted of bread, butter and tea for every meal.

However, a system of fines and deductions meant that taking a full wage packet home was uncommon. Fines of a few pence to a shilling were imposed on girls who dropped matches, or talked, or went to the toilet without permission. In addition, the girls were told "not to mind their fingers" around the factory machinery, even if it caused an injury, and were subject to "blows" from the foremen.

The white phosphorus that gave the matches their striking tip made working at the factory a deadly business. Exposure to the toxin caused phosphorus necrosis of the jaw, nicknamed "phossy jaw", which caused the jawbone to rot. Those suffering from phossy jaw would develop a terrible toothache and swelling of the gums; the jaw would take on an eerie green glow that could be seen in the dark. Surgery to remove the jaw was often the only option to save a sufferer of phossy jaw; brain damage and death from the condition were common.

The owners of Bryant and May, who collected a 22 per cent dividend from their business, refused to accept Besant's findings and even tried to force the matchstick girls to sign affidavits that the accusations were untrue. However, the workers had had enough and agreed to strike. In a pioneering victory for the British labour movement, Bryant and May were eventually forced to offer their workers a new deal including higher wages and the abolition of fines. The use of white phosphorus in British matchsticks continued until 1906, when it was finally banned. Instances of phossy jaw completely disappeared afterwards.

ABOVE: A Victorian worker shows the devastating effects of phossy jaw.

ABOVE: The matchstick girls from the Bryant and May factory in London's East End. Their successful strike was an early victory for the British Labour movement.

Phosphorus

Phosphorus is a non-metallic element found in phosphate rock and in trace amounts in human urine. It carries the chemical symbol P and comes in white, red and black forms.

OVERVIEW:

It has been suggested that phosphorus was first isolated by twelfth-century Arab alchemists, although it was not rediscovered until 1669. German alchemist Hennig Brand surmised that because phosphorus shone in the dark it was perhaps the legendary philosopher's stone. To extract phosphorus, Brand would leave 50 buckets of urine to stand until they "bred worms" before boiling the urine down into a paste. This was heated with sand to distil the phosphorus. In the 1800s, James Burgess Readman of Edinburgh developed a furnace that could produce phosphorus from phosphate rock, which is how it is extracted today. Because it was the 13th element discovered and was used in poisons, explosives and nerve gases, it was given the title "the devil's element". Used in fertilizer, matches and fireworks, phosphorus spontaneously combusts in air; it was used in British incendiary bombs of World War II and other conflicts since. White phosphorus burns fiercely. It can ignite combustibles such as fuel; it sticks to cloth, skin and human tissue and can cause third-degree burns. Nerve gases, the most toxic substances known to humans, are organic derivatives of phosphorus.

TOXIC EFFECTS:

Of its different forms, white phosphorus has traditionally been used as a poison. It is extremely toxic if ingested: 15 mg can be enough to kill an adult. Some survivors have reported excreting smoking stools after passing the poison. Its extreme toxicity is due to the fact that it contains free radicals that are not easily metabolized in the liver and build up to dangerous levels. This can lead to kidney, liver and heart damage, as the poison attacks the sufferer's central nervous system. It can also be inhaled – which leads to the condition known as "phossy jaw" (page 96) – or absorbed by the skin as an incendiary agent. When this occurs, third-degree burns, poisoning, and multiple organ failure can occur.

SYMPTOMS:

Symptoms of phosphorus poisoning after ingestion include abdominal pain, diarrhoea and vomiting of luminescent matter, followed by convulsions, coma and death. Death typically occurs between several hours and three days afterwards. The symptoms of inhalation of vapours or absorption through skin usually appear within two days and include nausea, coughing, vomiting, extreme fatigue, numbness, low blood pressure, heart problems, convulsions and death. Chronic symptoms of long-term inhalation include the rotting of the jaw, phossy jaw.

TREATMENT:

There is no specific antidote for phosphorus poisoning. Stomach flushing and activated charcoal can help to evacuate the poison, although inducing vomiting is not recommended. Surgical removal of the jawbone is required for phossy jaw.

FAMOUS POISONINGS:

- In the 2008–09 Gaza War, the Israeli military dropped incendiary shells filled with white phosphorus over civilian areas in Palestinian-occupied Gaza, including on a refugee camp and a school, where children were sheltering. More than 1,400 Palestinians were killed during the war, including from wounds where the phosphorus had burned through the flesh to the bone.

OPPOSITE: A lump of phosphate rock in its pure state.

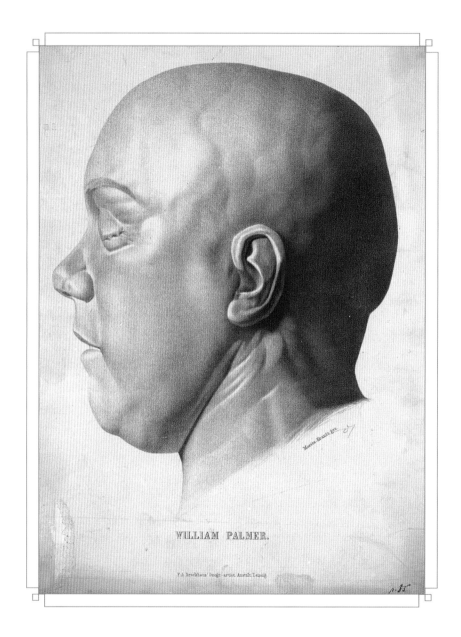

WILLIAM PALMER.

Villain William Palmer

Over 30,000 spectators crammed the streets of Stafford to see the execution of doctor William Palmer. Some had spent the night in the rain to secure their place, such was the notoriety of Palmer, the "Prince of Poisoners", who had murdered for money.

On 14 June 1856, Palmer was hanged; his noose was then cut into pieces and sold as souvenirs.

William Palmer poisoned for the first time after qualifying as a surgeon at St Bartholomew's Hospital, London. Palmer challenged the husband of a love interest to a drinking competition. When the man returned home, he collapsed and died; poison was suspected but not proven.

Soon afterwards, Palmer began a practice in his home town of Rugeley and married Ann Thornton, a young woman with a famously wealthy mother. Ann's mother lent money to Palmer and then mysteriously died after staying with the couple for two weeks. However, the mother had left the couple little money in her will, leaving Palmer disappointed. He began gambling heavily on the horses at this time.

Palmer's companion at the racetrack was one Leonard Bladen, who loaned the doctor hundreds of pounds over the space of a few months. Then, one night at Palmer's house, Bladen was seized with a terrible pain and died shortly afterwards. The money he had won that day at the track disappeared, and his cause of death was mysteriously noted as "injury of the pelvis".

More deaths closer to home followed. Palmer's wife Ann gave birth to five children, but four of them perished before they reached three months: one of Palmer's sons died three days after being born; another after seven hours. But infant fatalities were common at that time, and the cause of death for all four was listed as "convulsions". It is now thought Palmer dipped his finger in strychnine or antimony – his poisons of choice – followed by sugar to entice his babies to suckle it.

By reducing the number of mouths to feed, Palmer had lessened his outgoings, but he was still heavily in debt. His cycle of gambling, living a lavish lifestyle beyond his means, and then killing to cover his debts followed him until his arrest for murder. The next victim on Palmer's list was his 27-year-old wife Ann; he took out life insurance for her and then killed her a few months later. The cause of death was recorded as cholera, which killed thousands in Britain at that time. Now free of filial ties, Palmer was able to live the reckless life of a bachelor. However, the £13,000 paid out by Ann's insurance policy had not been enough to cover his debts.

With creditors threatening, Palmer took out a life insurance policy on his brother Walter, a notorious drunk. Palmer then took out another policy on a farm hand, an acquaintance of his. When both Walter and the farm hand died, the insurance companies became suspicious. Neither firm paid out, and an investigation was launched; foul play, however, could not be proven.

Palmer's luck would not last. By 1855, he was seriously in debt, his creditors were threatening to expose him to his mother – a source of his revenue – and he had just fathered an illegitimate child. Trying to recover his losses, Palmer spent much time at the track, accompanied by an independently wealthy friend called John Cook. While celebrating his winning at the races – Palmer had lost heavily as usual – Cook clutched at his throat and complained

OPPOSITE: An illustration of William Palmer's death mask, taken from a bust of the same. Palmer's head had been shaved prior to execution.

that his gin was burning it. He then became ill. This happened again in Palmer's presence, and then, after Palmer sent his friend a bottle of gin as a present, the man began vomiting. As a doctor, Palmer made himself available to treat Cook, which he did by administering three grains of strychnine. Cook immediately screamed in agony and shouted that he could not breathe; he died with his back violently arched.

Palmer then set about collecting Cook's outstanding race winnings on his behalf and doing his best to sabotage a post-mortem on his friend. He did this by bumping into the attending doctor as he worked and tampering with part of Cook's stomach collected into a jar as evidence. A second post-mortem was held, which failed to identify poison as the cause of death. However, the doctor responsible for the post-mortem made no bones about his suspicions, saying that the "deceased died of poison wilfully administered to him by William Palmer".

More conclusive evidence was to follow. The bodies of Palmer's wife Ann and his brother Walter were exhumed, and Ann's was found to contain high traces of antimony. There was other circumstantial evidence linking Palmer with the crime: two chemists testified that they had sold Palmer strychnine, but the defence countered that traces of that poison had not been found in Cook's or Ann's stomach. For a time, it looked as if Palmer might walk free, especially as the presiding judge told the jury that he believed the doctor to be innocent.

However, when presented with the prosecution's case that Palmer had murdered for money to avoid going to debtors' prison, the jury

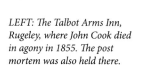

LEFT: *The Talbot Arms Inn, Rugeley, where John Cook died in agony in 1855. The post mortem was also held there.*

ABOVE: St. Augustine's Churchyard, Rugeley, where John Cook and members of the Palmer family are buried.

took notice. After one hour they returned the verdict of guilty, and Palmer was sentenced to be hanged. The sentence created great controversy as there had been little hard evidence to convict Palmer. However, after his execution damning titbits emerged, such as a prescription that Palmer had written himself for strychnine and opium.

On the hangman's scaffold Palmer was asked if he wanted to confess his crimes, which he didn't. After he was hanged and parts of the noose sold off for 5 shillings per inch, two death masks were made from Palmer's body. It was then deposited into a sack and buried in a mass grave underneath the prison chapel window. Contemporary author Charles Dickens called William Palmer the "the greatest villain who ever stood in the Old Bailey". Palmer's life of poisoning friends and family members for profit is his legacy.

ANTIMONY

Antimony is a silvery grey metallic element, which was once used in small doses to worm animals but is highly toxic for humans in large doses. A lethal dose of antimony is between 100 and 200 mg, and the poison has a toxic action similar to arsenic. Antimony inactivates key enzymes in the body and destroys red blood cells. Symptoms of acute antimony poisoning are gastroenteritis, nausea, vomiting and bloody diarrhoea followed by kidney depression. There is no antidote for antimony poisoning, so treatment can include having the stomach pumped and then administering the drug dimercaprol to hasten the excretion of the poison. A blood transfusion to replace destroyed red blood cells is also sometimes administered.

Strychnine

Strychnine is a poison obtained from the seeds of the nux vomica tree (S. nux-vomica) and plants from the Strychnos family.

OVERVIEW:

First discovered in the Saint Ignatius' bean (*S. ignatii*) by French chemists in 1818, strychnine is perhaps best known for its appearances in Agatha Christie's murder-mystery novels. In Christie's day, strychnine was easily obtainable over the counter in chemists' shops, both as a rat poison and as a medicine. Christie characters such as Mrs Inglethorp were prescribed strychnine as a nerve tonic and appetite invigorator; its use as a poison was probably introduced because of its dramatic effects, which include convulsions and asphyxiation. Most recorded incidents of people being murdered with strychnine come from the nineteenth century, although they are still rare compared to other poisons. However, strychnine has been widely used for centuries as a poison to kill animals. It was also used in lower dosages in the late nineteenth century as a performance-enhancer in sports because it had similar effects to coffee, giving the consumer greater physical endurance and a restless, energized feeling.

TOXIC EFFECTS:

Strychnine affects a person's ability to control movement. It does this by binding with one of the two neurotransmitters that send messages between nerve cells. One of these neurotransmitters (acetylcholine) tells a nerve to fire; another (glycine) tells it to stop. Strychnine attaches to glycine receptors and therefore blocks the signal for a nerve to stop. The result is the continuation of a nerve's movement, which is increased by the slightest stimulus. This causes a person to twitch uncontrollably, followed by the violent convulsions often associated with the fictional victims of murder mysteries.

SYMPTOMS:

Strychnine attacks the central nervous system and causes the victim's muscles to contract simultaneously, especially in the legs, arms, back, neck and face. This can cause the back to arch violently and facial muscles to become taut, causing the victim to smile as they die. Death is generally caused by suffocation as the strychnine eventually stops a person's ability to breathe. However, the victim remains conscious as this takes place because the poison also stimulates the brain, giving it a heightened sense of perception. In the end, the body is left still in a type of frozen spasm with the eyes wide open. It is thought to be an agonizing death.

TREATMENT:

It is possible to avert death through strychnine poisoning if the stomach is flushed before the symptoms appear. Once the symptoms do appear, it is possible to limit the severity of the convulsions through muscle relaxants, tranquillizers and rest in a dark, quiet room without stimuli. If the symptoms are kept under control, the poison can pass through the body in around 24 hours.

FAMOUS POISONINGS:

- Nazi SS officer Oskar Dirlewanger was a notorious sadist and psychopath who used to strip Jewish women prisoners and inject them with strychnine to watch them convulse to death for entertainment.

OPPOSITE: *An illustration of the* nux vomica *tree.*

Thomas Cream's Chloroform

As Dr Thomas Neill Cream was hanged at Newgate Prison in 1892, he reportedly uttered some last words that were interrupted by the noose: "I am Jack …". It was a shocking confession.

Was it possible that Cream, the "Lambeth Poisoner" who had been sentenced to death for the murder of prostitutes, was also Jack the Ripper? With Cream, a sadist who had killed multiple women on two continents, anything seemed possible.

Born in Glasgow in 1850, Cream was raised in Canada and graduated with a medical degree from Montreal's McGill College. His thesis had been on chloroform. Cream used chloroform to perform an abortion on Flora Brooks, a woman he had made pregnant and then nearly killed through the illegal operation. Brooks's enraged father insisted that the two marry, which Cream did before departing for post-graduate medical schools in London and Edinburgh. Cream left behind some "medicine" for Flora, who died soon afterwards – allegedly from consumption, although no autopsy was performed.

After his return to Canada in 1879, Cream was accused of the murder of Kate Gardener, his patient and mistress. Gardner was found dead in a shed behind his office, pregnant and poisoned by chloroform.

Cream fled to Chicago, where he opened a practice performing illegal abortions for prostitutes. In 1880 he was accused of the murder of prostitute Mary Ann Faulkner, found dead in Cream's assistant's apartment following an abortion. Cream was found not guilty after blaming his assistant for the botched abortion. However, he would soon get his comeuppance.

After advertising in a Chicago newspaper anti-epilepsy pills that he had invented, Cream poisoned one of his patients, Daniel Stott. He then convinced Stott's wife, whom he had seduced, to blame the chemist for administering the wrong pills. Cream even telegraphed the coroner advising him of this. This alerted the coroner not to the possibility of an accidental poisoning, but to that of foul play. He tested Cream's pills by feeding them to a dog; it dropped dead 15 minutes later. The coroner then ordered the exhumation of Daniel Stott, whose stomach was found to contain enough strychnine to kill three men.

The trail led directly to Cream, whose lover, Julia Stott, exposed him on the witness stand. Once again Cream fled, but this time he was arrested and jailed for life for murder. After 10 years Cream's father died, leaving his sons wealthy men. Cream's life sentence was mysteriously commuted soon afterwards. In 1891, Cream was set free.

The sinister, foul-mouthed drug-addict who emerged from Illinois' Joliet prison was a stark contrast to the dapper doctor of a decade earlier. He now regularly ingested morphine, cocaine and strychnine and spoke of his sadistic hatred of women. He would act out this hatred after sailing to London.

Cream began prowling the city's East End slums for victims. According to a later police report, Cream was in the habit of accosting strangers with wild, rambling talk and showing them pornographic

OPPOSITE: Doctor Thomas Cream was a murderer of women on both sides of the Atlantic.

pictures, which he carried with him. "Women were his preoccupation and talk of them was far from agreeable," the report noted.

Now going by the name "Dr Neill", Cream began poisoning the street prostitutes of the impoverished London borough of Lambeth. His first victim was 19-year-old Ellen Donworth, who reported an encounter with a "topper" (slang for gentleman with a top hat) on 13 October 1891; she shared a drink with him and died of strychnine poisoning two days later. Cream then made the extraordinary move of writing to the coroner offering to expose the murderer for £300,000.

Next was the prostitute Matilda Clover, who died the day after meeting with Cream. Although her death was officially diagnosed as alcohol poisoning, Cream nonetheless wrote to a local doctor accusing him of Clover's poisoning and demanding money for his silence. The doctor handed the letter to the police.

After the murders, Cream returned to visit relatives in Canada, where he frequented brothels and procured a large quantity of strychnine pills. On the boat back to London, Cream alienated himself from the other passengers by getting drunk on whisky and boasting about the cheapness of London prostitutes, who he said could be hired for "a shilling each".

Back in London in April 1892, Cream attempted to murder prostitute Lou Harvey, but the attempt failed after she threw the pills he gave her into the River Thames. He then killed two prostitutes, Alice Marsh and Emma Shrivell, by giving them a bottle of Guinness laced with strychnine.

Cream's next meeting was not with a prostitute, but with a New York detective. The detective had read about the murders by the so-called "Lambeth Poisoner" and found that the doctor, whom he met by chance, had an encyclopaedic knowledge of every case. Cream even took the detective to various sites of interest around Lambeth, which was enough for the detective to suggest to his colleagues at Scotland Yard that Cream might be the killer.

The detectives of Scotland Yard had been trying to track an anonymous blackmailer who had written letters demanding money from prominent people,

WAS CREAM JACK?

Theories that Thomas Cream was in fact the notorious serial killer, Jack the Ripper, made for sensational headlines. Does the claim have any truth? The uniting factor is that both men were misogynists who operated in London's East End and enjoyed murdering prostitutes. However, during Jack the Ripper's most prolific period in 1888 Cream was held in Joliet prison, giving the doctor a cast-iron alibi. However, conspiracy theorists have proposed that Cream bribed prison guards to have him replaced with a look-alike and travelled to London several years earlier than officially recorded. It remains an unproven theory.

LEFT: Elizabeth Stride was a Jack the Ripper victim who had fallen into prostitution after her marriage failed. Her throat had been slit.

> **DESCRIPTION OF CREAM**
>
> This account of Dr Thomas Cream was published in the *Canadian Medical Association Journal* after his death:
> "He [Cream] was a drug fiend and this may have been a factor in his career of habitual murder. His lustful habits and the debasement of moral fibre through drugs led him to the point of losing all sense of moral obligation. He became a sensualist, a sadist, drug sodden and remorseless, a degenerate of filthy desires and practices, who used his medical knowledge to slay his unfortunate victims. What suffering would have been avoided had not soft-hearted, misguided enthusiasts sought his release from Joliet!"

whom he had threatened with exposure as the Lambeth Poisoner. This, of course, was Cream, who had been stupid enough to include the "murder" of Matilda Clover in his letters, despite the fact that her death was officially recorded as occurring from alcoholism.

In the end it all led to Cream: the letters, the visits to prostitutes and previous jail time for murder by poisoning in the United States. In October 1892, Cream was sentenced to be hanged for the murder of the four prostitutes. Before his sentence, the coroner had read out a letter supposedly from Jack the Ripper. It said that Cream was innocent and should be freed. Cream laughed hysterically at this. So what of Cream's claim of "I am Jack …"? One theory was that as Cream lost control of his bodily functions – a side effect of being hanged – he cried "I am ejaculating", which was misheard.

ABOVE: East End bobbies shine a light on another murdered victim of Jack the Ripper.

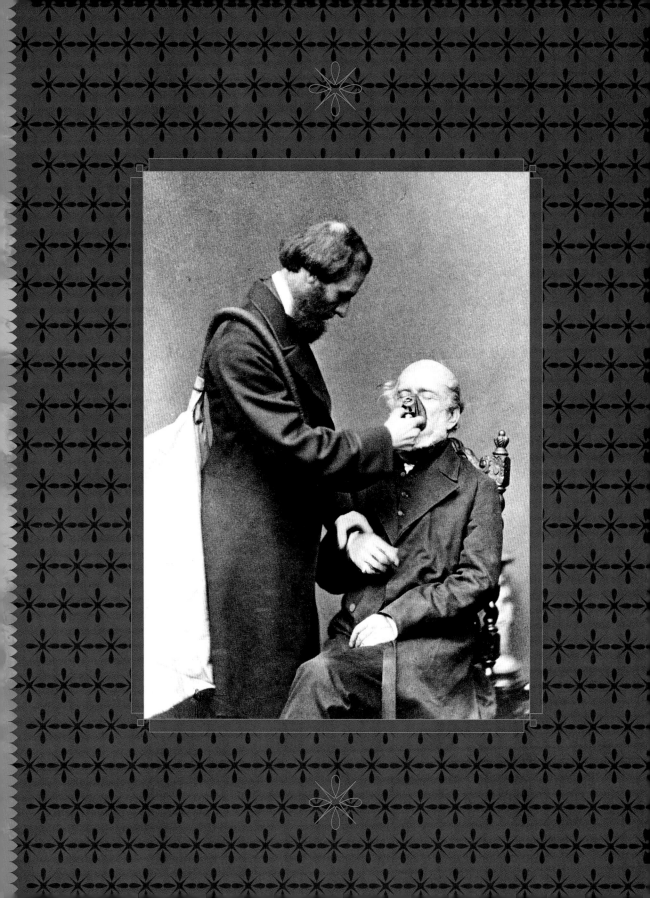

Chloroform

Chloroform is a clear, colourless liquid prepared by the chlorination of methane. Also known as trichloromethane, chloroform is most commonly used as an anaesthetic and sedative.

OVERVIEW:
Chloroform was discovered independently in the United States, France and Germany in the early 1830s; the first reports of its abuse followed soon afterwards. Dentist Horace Wells, the first American to use nitrous oxide in surgery, became addicted to chloroform, which was said to have completely changed his personality. He committed suicide in jail in 1848 after throwing acid at a passer-by in New York City. The anaesthetic qualities of chloroform were first tested in Scotland 1847 on a patient during surgery, and Queen Victoria gave birth to her last two children under its influence in the 1850s.

However, the dark side of chloroform had also emerged: a 15-year-old girl died from chloroform poisoning in 1848 after having it administered to remove an infected toenail. Others abused chloroform as an intoxicant or a method of suicide. By the twentieth century, criminals had discovered that chloroform could render their victims unconscious. Chloroform in higher doses can easily kill a grown adult. Its use as an anaesthestic diminished after the 1930s, when less toxic alternatives became popular.

TOXIC EFFECTS:
Chloroform can be absorbed through the lungs, gastrointestinal tract and even the skin. Once ingested, chloroform is rapidly absorbed and distributed to all organs. This can result in central nervous system depression, changes in the respiratory rate, heart irregularities, intestinal pain and toxicity in the kidneys and liver. Chronic exposure can lead to serious blood, kidney and liver damage. Even 10 ml of chloroform is enough to kill an adult.

SYMPTOMS:
The symptoms of chloroform poisoning include headaches, nausea, vomiting, irritability, confusion, drowsiness and intoxication, followed by respiratory arrest, heart attack, coma and death. Unconsciousness caused by chloroform usually takes several minutes depending on the dose. The difficulty in calculating a safe dose and chloroform's toxic effects on the organs are two of the reasons why it is no longer favoured.

TREATMENT:
Eyes and skin that have come into contact with chloroform can be treated with warm water and saline solution. If inhaled, oxygen can be administered, and if ingested, the drug N-acetylcysteine (NAC) can help prevent kidney and liver damage and death.

FAMOUS POISONINGS:
- American H. H. Holmes was a nineteenth-century swindler and con-man known as America's first serial killer. He would use chloroform on his victims to trap or kill them in his purpose-built murder castle before cashing in their life insurances.
- In 1900, American philanthropist William Marsh Rice was murdered by his valet, who was plotting with Rice's lawyer to steal his fortune. The valet killed Rice by administering chloroform while he slept, but foul play was later suspected when a large cheque made out to Rice's lawyer was presented at the bank. Rice's money was used instead to found Rice University in Texas.

OPPOSITE: *Doctor Joseph T. Clover administers chloroform with his 'Clover's chloroform apparatus'.*

Chapter 5: Twentieth-Century Poisons

In the twentieth century, the nature of poisoning changed. Government regulations banning the public sale of toxic substances such as arsenic made it harder to buy traditional poisons over the counter.

Forensic toxicology was also becoming an effective scientific discipline, relegating the craze of great arsenic poisonings of the nineteenth century to the history books.

However, alongside the desire of governments to regulate the sale of poisons came the manufacture, often led by governments, of toxic substances on an industrial scale. Many of these were developed for the common good, such as pesticides and herbicides to kill harmful insects and vermin and increase the yield of crops. However, in secret government laboratories, these substances were found also to contain extremely noxious poisons capable of wiping out entire cities if delivered correctly.

Industrial poisons for use in war were soon being manufactured for use in global conflicts. In 1916, sulphur mustard, aka mustard gas, was developed by the German military and used to terrible effect in the trenches of World War I. The blistering effects of the gas left its victims with first-degree burns, disfiguring scarring, blindness and a greater chance of cancer if they survived.

The German manufacture of industrial poisons once again emerged during World War II. This time a form of hydrogen cyanide called Zyklon B was used in Nazi extermination camps to murder over one million people. At the end of the war the Nazi high command poisoned themselves with cyanide rather than risk capture by the Allies.

In 1946, a group of Jewish guerrillas from the Vilna ghetto in Lithuania smuggled arsenic into Germany to poison their former Nazi prison guards who were awaiting trial. The Jews painted 3,000 loaves of black bread with arsenic bound for the prisoners. The plot largely failed, but over 2,200 were taken ill.

Poisoners became mass murderers in the twentieth century; industrial-scale production of toxins enabled them to practise large-scale "retribution" against their victims. In 1978, American Reverend Jim Jones ordered the mass suicide of over 900 of his followers in Jonestown, Guyana; in 1995, cult leader Shoko Asahara unleashed several sarin packets on the Tokyo subway during rush-hour one morning.

Sarin was one of the Cold War poisons developed in secret weapons facilities across Europe. However, it was not until the twenty-first century that the deadliest poisons created in these facilities emerged. These military-grade nerve agents and radioactive toxins would become household names in the great geo-political poisonings of modern times.

OPPOSITE: German soldiers pose with gas masks and grenades in their trench. Chlorine, phosgene and mustard gas were the three main chemical weapons of World War I.

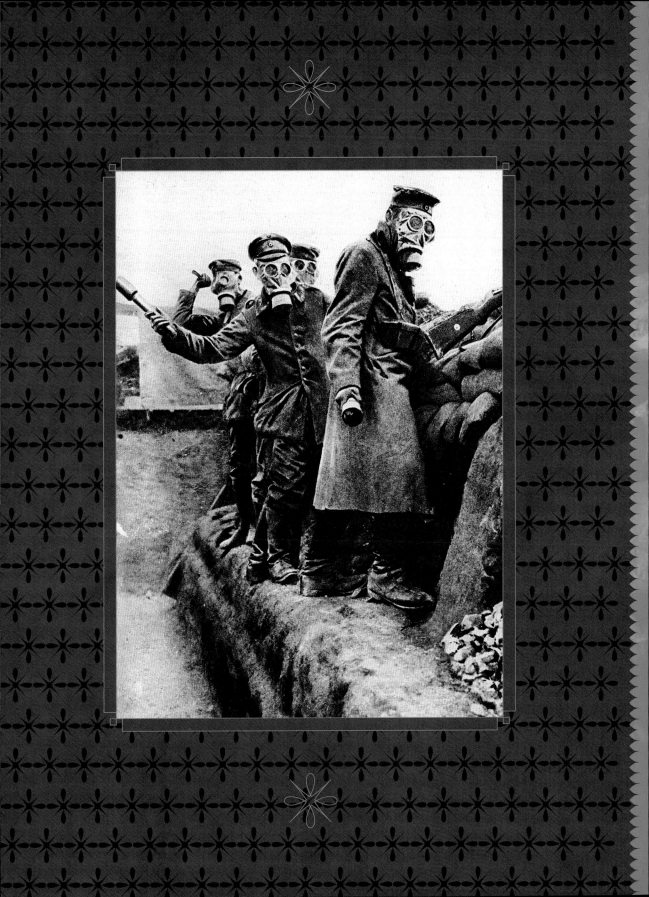

Assassinating Grigori Rasputin

Grigori Rasputin is best known to history as Russia's "Mad Monk", a mystic with an insatiable appetite for sex and alcohol who ingratiated himself with the royal family and survived several assassination attempts on his life.

The lurid details of Rasputin's murder remain a highly controversial topic to this day: it was said that even cyanide was not enough to finish off this unkillable man.

Born in 1869, Rasputin was an illiterate Siberian peasant who became a wandering mystic and holy man and later a prominent member of the Tsar's inner circle. He was introduced to Tsar Nicholas and his wife Alexandra by Russian Orthodox clergymen, and made an immediate impression on the royal couple. A five-minute introduction lasted for an hour, and Rasputin was soon invited to the royal palace.

There, Rasputin made a considerable impression by "curing" royal prince Alexi of his haemophilia, apparently through prayer. It is thought that the miracle actually occurred because Rasputin banned all doctors and thus the aspirin, a blood-thinner that was widely prescribed as a cure-all at the time. Afterwards, Rasputin became an indispensable member of the royal entourage.

Rasputin's influence over the royals, however, irked many high-ranking Russian officials. He was rumoured to be providing political advice to the Tsar, sleeping with Alexandra, and being behind a treasonous plot to cause a cholera outbreak in St Petersburg with poisoned apples. Rasputin's sexual appetites for women from all classes were legend, although Alexandra dismissed his infidelities by saying, "He has enough for all."

This was all too much for Felix Yusupov, husband to Tsar Nicholas's niece. Yusupov decided that Rasputin was a menace who must be destroyed. To this end, he invited Rasputin to his palace to meet his wife. According to Yusupov's memoir, the only written account of his death, Rasputin was served with a platter of cakes that had been laced with potassium cyanide. He also consumed a number of glasses of wine spiked with cyanide. To Yusupov's amazement, the poison appeared to have no effect on Rasputin.

Yusupov then shot Rasputin several times with a revolver. But afterwards Rasputin's "green eyes of a viper" flickered open and he attacked Yusupov. Yusupov and his co-conspirators then chased Rasputin outside the house and clubbed him to death. His body was discovered in the Neva River. Responses to Rasputin's death were mixed: he was mourned by peasants and the royal family, but the Bolsheviks saw him as a symbol of the imperial court's corruption. The Bolsheviks overthrew the monarchy in 1917 and killed the royal family soon afterwards.

ABOVE: Rasputin's body was dragged battered and frozen from the Malaya Nevka River.

OPPOSITE: Rasputin is surrounded by some of his female acolytes: his many affairs were legend.

Cyanide

Cyanide is a dangerous toxin that exists in several forms. As a poison, cyanide is typically the colourless gas hydrogen cyanide (HCN), or the crystalline sodium cyanide (NaCN) and potassium cyanide (KCN). Cyanide occurs naturally in cassava and fruit seeds and pips, such as apple, apricot and peach; it is also emitted from fire and cigarette smoke and used in the manufacture of synthetic materials.

OVERVIEW:
Cyanide is considered one of the big three poisons alongside arsenic and strychnine. The ancient Egyptians knew of its existence and recorded it as "death through peach stones". Cyanide is at its most lethal when used as a gas, and the hydrogen cyanide compound Zyklon B was used to murder over a million prisoners in the Nazi extermination camps in World War II. Then, inmates were told to undress to shower for lice and driven into purpose-built gas chambers disguised as communal showers. Zyklon B pellets were then dropped into the chambers via a small hole in the roof. Exposure to moisture and human body heat caused the pellets to give off the hydrogen cyanide gas. It took around an hour for all in the chamber to die. Nazi leaders in Berlin committed suicide with cyanide capsules rather than be caught by the approaching Soviet Red Army in 1945. In 1978, over 900 members of the Jonestown community, Guyana, committed mass suicide by drinking Flavor Aid laced with cyanide.

TOXIC EFFECTS:
Cyanide poisoning works by inhibiting the oxidative processes of human cells. It does this by preventing red blood cells from absorbing oxygen by interfering with the body's enzymes. When the cells are starved of oxygen, the process of chemical strangulation begins. This often occurs so quickly that it causes death before any of the other symptoms have time to appear.

SYMPTOMS:
The symptoms of acute cyanide poisoning are dizziness, headache, nausea, slow heart rate, loss of consciousness, convulsions, lung damage, respiratory failure, confusion, coma and death. A lethal dose of cyanide can be as little as 300 mg of solid cyanide or 100 mg of hydrogen cyanide. Death can occur within 20 minutes of ingestion or inhalation.

TREATMENT:
Because cyanide works so quickly, effective treatment depends on the speed at which an antidote is delivered. Cyanide antidote kits are available and these include the drugs amyl nitrate, sodium nitrate and sodium thiosulfate.

FAMOUS POISONINGS:
- In the late 1950s, KGB agent Bohdan Stashynsky assassinated Ukrainian nationalist leaders Lev Rebet and Stephan Bandera with a spray gun that fired gas from a crushed cyanide capsule.
- In 1982, seven people died in Chicago from packets of the painkiller Tylenol spiked with cyanide. The poisoner was never found, although the incident did lead to tamper-proof packaging.
- In 2013, poachers in Zimbabwe killed over 300 elephants and other safari animals after dumping cyanide used in gold-mining into a watering hole. Several humans then died after eating the contaminated meat.

OPPOSITE: The crematorium building at the German Majdanek extermination camp, Poland.

Hitler's Cyanide Command

In Adolf Hitler's bunker on 30 April 1945, news came through the telephone switchboard. The German army had failed to break though the Soviet encirclement of Berlin.

The end of Hitler's so-called "1,000-year Reich" and World War II were now inevitable. Rather than be captured by the enemy, Hitler, his wife Eva Braun and members of his high command used cyanide to kill themselves.

After hearing the news of the German army's failure, Hitler spoke quietly with head of the Nazi party Chancellery Martin Bormann and shook hands with his adjutant Otto Günsche. Hitler told Günsche all German soldiers were now released from their oath of loyalty. Hitler and his wife Eva Braun shut the door to his study, and Günsche told those outside not to disturb them.

Everyone in the bunker fell silent and waited nervously. Finally, there was some commotion. The study door was opened by bodyguard Rochus Misch to reveal the scene inside: "My glance fell first on Eva. She was seated with her legs drawn up, her head inclined towards Hitler. Her shoes were under the sofa. Near her … the dead Hitler. His eyes were open and staring, his head had fallen forward slightly."

Hitler and Braun had both bitten down on glass cyanide ampoules, and Hitler had also shot himself in the head. Their bodies were taken out of the back door of the *Führerbunker* and set alight with petrol as Hitler had instructed his valet Heinz Linge: "You must never allow my corpse to fall into the hands of the Russians. They would make a spectacle in Moscow out of my body and put it in waxworks."

With Hitler gone, members of his high command also began killing themselves. The new chancellor Joseph Goebbels and his wife Magda shut their six children in a bedroom, gave them a morphine injection and forced their teeth together on glass cyanide ampoules. Magda then came out of the room crying and sat down at a table to play patience. On 1 May, the couple walked out of the bunker, bit into cyanide ampoules and shot themselves.

Dozens more high-ranking Nazis and prominent German generals killed themselves in the last days of World War II. Heinrich Himmler, the main architect of the Holocaust, and Nazi party leader Hermann Goering both committed suicide with cyanide after being captured by the allies.

On 3 May, soldiers from the Soviet army entered the *Führerbunker* and discovered the dead bodies of the six Goebbels children. Their faces showed the gruesome evidence of cyanide poisoning. The nearly one million dead bodies killed by cyanide poisoning in the Nazi extermination camps had also been recently discovered.

ABOVE: *The front page of New York's* Daily News, *May 2, 1945.*

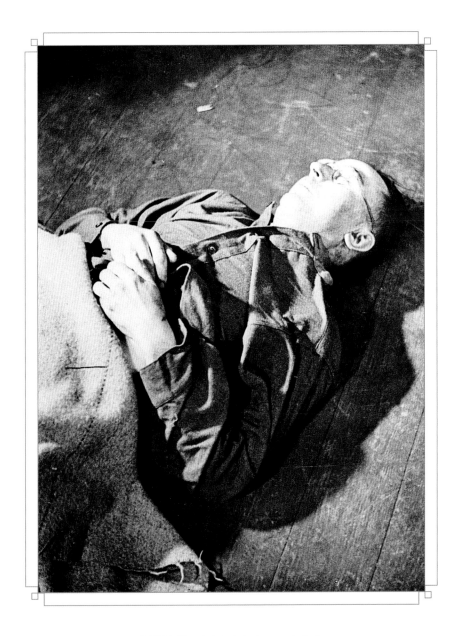

ABOVE: Himmler's body is photographed after he committed suicide using a cyanide pill. It took fifteen minutes for him to die.

Alan Turing Tragedy

On 8 June 1954, Alan Turing was found dead in his bed. Turing, a mathematical genius, cracker of the Enigma code, and pioneer of computer science, was found to have committed suicide by ingesting cyanide.

A half-eaten apple coated with the poison was apparently discovered next to his body. However, some believe that instead of suicide, Turing died as the result of an accident, or even foul play.

In the early 1950s, Alan Turing was a respected mathematician undertaking ground-breaking work at Manchester University on the theory of artificial intelligence. Few of his colleagues would have dreamed that Turing was the man responsible for inventing the British Bombe: the revolutionary machine that cracked Enigma, the coding machine used by the Nazis to send battle orders during World War II. The Enigma code was believed to be impossible to crack until Turing created the Bombe. This enabled the Allies to intercept messages during many subsequent battles, including the pivotal Battle of the Atlantic. Cracking Enigma was said to have ended the war two years early and saved millions of lives.

Turing was a war hero, but his work on cracking Enigma was highly classified; he could not talk about it. In addition, Turing was already considered to be a security risk because he was homosexual. Many in the intelligence services thought Turing could be lured into a homosexual honey trap and blackmailed into giving out state secrets, as homosexuality was a crime at that time.

Turing had been exposed as gay when his house was burgled and he explained to police he thought an acquaintance of his male lover might have been responsible. Astonished by the admission, the police arrested Turing and charged him with gross indecency. As an alternative to prison time, Turing accepted hormone therapy designed to reduce libido. This involved being injected with oestrogen over the course of a year. The therapy caused Turing to gain weight, grow breasts and become impotent.

After the conclusion of the treatment, it became known to security services that Turing was continuing to seek out gay lovers, including in countries behind the Iron Curtain. Did the intelligence services kill Turing and dress it up as suicide? Was Turing depressed enough by his conviction and treatment to kill himself? Or did he make a mistake with one of the many experiments he undertook in his room with cyanide? Whatever the cause, Turing's death was a tragedy that saw the loss of one of Britain's greatest minds. In 2017, the "Alan Turing Law" retroactively pardoned all men prosecuted for the now-abolished crime of homosexual activity.

ABOVE: *The back of Alan Turing's enigma-breaking Bombe.*

OPPOSITE: *Mathematician Alan Turing.*

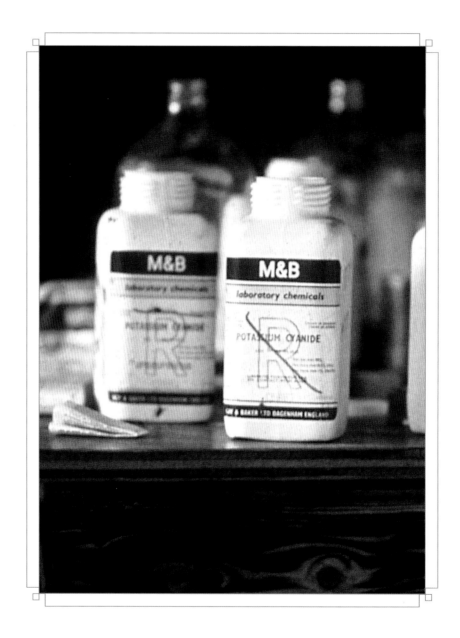

ABOVE: Some of the many bottles of potassium cyanide discovered at Jonestown.

The Jonestown Massacre

In November 1978, followers of Reverend Jim Jones' Peoples Temple were called to an emergency "White Night" in Jonestown, Guyana.

They were given a red liquid to drink and told it contained a poison that would kill them in 45 minutes. When no one died, Jones explained it was just a rehearsal. However, the largest mass-suicide by poisoning in American history was soon to follow.

Preacher Jim Jones set up his first Peoples Temple in Indianapolis in 1955. His church sought to create a new socialist community that had multi-racialism at its heart, a highly progressive notion during the era of American segregation. Thousands of followers flocked to the charismatic Jones, who appeared to read minds and heal the disabled during his sermons. By 1973, Jones had established Peoples Temples in San Francisco and Los Angeles and had become a respected churchman among politicians and the press. Jones also built a settlement for his followers outside the Californian town of Ukiah, one of the few places Jones believed would avoid the fallout of an imminent nuclear holocaust.

While Jones enjoyed popularity in public, he was a different figure behind the closed doors of his Temple communities. Here, followers were beaten, blackmailed, ritually humiliated and coerced into signing their properties and money over to Jones. Family members were encouraged to spy on each other and black followers brainwashed into believing they would be forced into government concentration camps if they defected from the Temple. When Jones faced accusations of financial fraud and sexual and physical abuse of his followers, he moved his temple to Guyana to build a socialist utopia called Jonestown.

Around 1,000 Peoples Temple members migrated from the United States to the Jonestown commune in 1977. Here, in the middle of remote jungle, members toiled on the land from 6.30 am to 6 pm six days a week before attending Jones' evening lectures on socialism, revolution and Temple enemies. These typically included Jones' former followers who had defected and the 'capitalist' US government. Although Jones had several million dollars in a bank account, members of Jonestown suffered from malnourishment and illness. Reports that some members were being held against their will filtered back to the American press and government.

ABOVE: Jim Jones was a Pentecostal preacher who described himself as a prophet and became increasingly obsessed with power and control.

By 1978, Jones had developed a serious addiction to pharmaceutical drugs and become obsessed with the idea that the CIA would mount a violent raid on Jonestown. He began enacting White Night emergencies, rehearsals for a "revolutionary suicide" – the mass-poisoning of Jonestown's residents by drinking cyanide mixed with the powdered fruit juice, Flavor Aid.

THE FLAVOR AID

The Jonestown Flavor Aid contained the poisons potassium cyanide and potassium chloride, and the sedatives valium, phenergan and chloral hydrate. Potassium chloride causes cardiac arrest and is used in lethal injections in US prisons. The sedatives were almost certainly used to mitigate the violent effects of the cyanide, which causes a painful death that resembles suffocation.

After being ingested, cyanide prevents the victim's cells from using oxygen to make energy molecules, which causes the cells to starve. This leads to cardiac and respiratory failure, during which the victim can only take irregular breaths. Limb spasms and body convulsions follow and the victim's face often contorts into a deadly grin called 'cyanide rictus'. The autopsies on seven Jonestown bodies did not find evidence of cyanide rictus, although this does not mean their passing was peaceful. In an audio recording of Jim Jones' final sermon, dying screams of pain are audible in the background.

ABOVE: Authorities rush to collect over 900 Jonestown bodies found face-down and decomposing in the jungle heat.

In November 1978, US congressman Leo Ryan led a delegation of journalists and Temple defectors called "Concerned Relatives" to Guyana to investigate Jonestown. Jones ordered the meeting to be rehearsed with live music and festivities, but on the day the leader cut a desolate figure. Clearly suffering from addiction and physical ailments, Jones ranted to newsman Tim Reiterman about conspiracies and martyrdom. "It was shocking to see his glazed eyes and festering paranoia face to face, to realize that nearly a thousand lives, ours included, were in his hands," Reiterman later wrote.

The following day, Ryan left Jonestown with several defectors and was attacked as he left the compound. The attack continued at a nearby airstrip, as armed Jonestown members riding a tractor began firing on the congressman's chartered airplanes. Five people including Congressman Ryan were killed. Meanwhile, back at Jonestown, Jones gathered his followers in the compound's pavilion to tell them that the congressman was to be killed and a violent government raid on Jonestown would follow. He then gave the order for his followers to commit "revolutionary suicide."

A vat of Flavor Aid was mixed with cyanide and sedatives as Jones sermonized through a loudspeaker. Jones ordered the concoction to be syringed into the mouths of the children first and advised his followers to "die with a degree of dignity…I don't care how many screams you hear." The adults then drank the poison from plastic cups, while puncture marks found in the arms of others suggest they were forcibly injected. Guards armed with crossbows and guns surrounded the pavilion and ensured Jones' orders were carried out; the preacher himself died from a gun-shot wound.

Over 900 people died at Jonestown on 18 November 1978, in the largest loss of American civilian life in a non-natural disaster before the terrorist attacks of September 11, 2001. Over 300 of the victims were children. While the view has been that the deaths were a mass-suicide by poisoning, many have argued that it was in fact mass-murder caused by the brainwashing and coercion of Jim Jones, the man known to his followers as "Father".

POTASSIUM CYANIDE

Cyanide is a chemical compound consisting of a bond of carbon and nitrogen (CN) that takes various forms, such as hydrogen cyanide gas and the crystalline solids sodium cyanide and potassium cyanide. Jim Jones had obtained many shipments of potassium cyanide after obtaining a jeweller's licence and claiming he needed the chemical to clean gold. Jones then ordered a Jonestown doctor to test the effectiveness of the poison on pigs. A memo from the doctor suggested around two grams would kill a large pig. Even more lethal than potassium cyanide is hydrogen cyanide, which was the main ingredient in the suicide pills taken by the Nazi high command. A high dose of hydrogen cyanide can kill an adult human in seconds.

ABOVE: An aerial photo of the bodies surrounding the central pavilion at the Jonestown compound, Guyana.

CRIPPEN.

THE ARREST.

COLLAPSE OF MISS LE NEVE.

INSPECTOR DEW'S DISGUISE.

DRAMATIC SCENE.

CAPTAIN KENDALL'S FULL NARRATIVE.

Dr. H. H. Crippen and Miss Ethel Le Neve were arrested in the Montrose at 9.30 yesterday morning (2.30 p.m. Greenwich time). They were described in the police notice as, "Wanted for murder and mutilation."

Scotland Yard at 4.5 p.m. received the following message from Inspector Dew, who had formally identified them:—

Crippen and Le Neve arrested. Will wire later.—Dew.

This wireless pursuit of Crippen is due alone to the acumen, astuteness, and ability of Captain Kendall, of the Montrose, whose exclusive messages to the "Daily Mail" have been a triumph of detective journalism.

Inspector Dew and the Canadian police were disguised as pilots when they boarded the vessel. Crippen betrayed anxiety as the boat approached, but was taken quite unawares when the police accosted him. Miss Le Neve almost collapsed.

Both were subjected to a lengthy examination by Mr. Dew, and it is understood that Crippen admitted his identity, and said that he was glad that the suspense was over. Several diamond rings were found in his possession.

He is charged with the murder and mutilation of his second wife, Mrs. Cora Crippen, known as Belle Elmore on the music-hall stage. The circumstances of the case are fully told on the next page. The Montrose carried him and the police on to Quebec yesterday. According to cablegrams, he will be sent back immediately.

THE ARREST.

(From Our Special Correspondent.)
FATHER POINT (viâ Rimouski),
Sunday Afternoon.

The long arm of British law reached its goal at half-past nine this morning, when two miles out in the River St. Lawrence Inspector Dew, of Scotland Yard, disguised as a pilot, pointed his finger confidently at a little man pacing the deck of the steamer Montrose, to see that he was having a difficult time over his rôle as pilot, and itched to assert himself as an officer of the law. Captain Kendall, Chief Constable McCarthy, and Inspector Dew chatted at the companion way. Detective Denis and Gaudreau turned forward to the wheelhouse.

Dr. Stewart and "Robinson" were walking up the deck. "Robinson" passed so close to Mr. Dew that the latter could have touched him. Still not a move was made. Inspector Dew was sizing up his quarry carefully—pitilessly. There could be no mistake. "Robinson" coughed slightly and turned towards the captain as though to ask a question. He was perfectly unconscious of the true state of affairs.

"Captain ——" he said almost jovially, tilting his grey hat to the back of his head. But that was all. His face became a blank; his knees shook together; and his arms went up as though to protect himself.

"I want to see you below a moment!" said Mr. Dew, with his characteristic lisp.

Then turning to Chief Constable McCarthy, he said, "That is the man!"

"I arrest you in the name of the King!" said Mr. McCarthy. "You are my prisoner! Anything you say will be taken down in writing and will be used against you at your trial!"

The passengers and crew, knowing for the first time that something out of the ordinary was going on, commenced to collect. Mr. McCarthy hustled his prisoner, not unkindly, down below.

As they were descending the narrow ship's stairs, Crippen said, "Have you a warrant? What is the charge?" Mr. McCarthy produced his authorisation for making the arrest—given him by Judge Angers, of Quebec.

Crippen grasped it before the Chief of Police could prevent him, and read it greedily. "Murder and mutilation!" he muttered to himself. "Oh, God!"

He threw the warrant on the floor of the passage, and walked to his cabin absolutely passive.

A few seconds later a woman's shriek told those above that Miss Le Neve had been discovered and arrested. She had recognised Inspector Dew in the semi-darkness of the passage, as she was emerging from her cabin to join Crippen.

When Mr. McCarthy entered he found her lying on the bed, fully dressed in boy's clothing. Her limbs were trembling and her face was as white as death. Mr. McCarthy said afterwards he thought she would break down immediately, but she recovered herself wonderfully, and when Inspector Dew came into the cabin she was quite composed.

As the pilot's boat swung away from the Montrose, Inspector Dew, Captain Kendall, Chief Constable McCarthy, and the two prisoners were closeted in the captain's cabin.

"GLAD SUSPENSE IS OVER."

THE "DAILY MAIL'S" WIRELESS MESSAGES.

CAPTAIN KENDALL'S NARRATIVE.

EXCLUSIVE TELEGRAMS.

The wireless telegram to the "Daily Mail" from Captain Kendall, of the Montrose, which appeared in our issue of Saturday, was the only wireless message sent by Captain Kendall to any newspaper.

On Saturday, July 23, the "Daily Mail" was the only newspaper to know that persons resembling Dr. Crippen and Miss Le Neve were on board the steamship Montrose. A wireless telegram was then despatched from London to Captain Kendall, asking him if he had Dr. Crippen and Miss Le Neve on board, how he had established their identity, and if they knew they were suspected. This was transmitted by the Marconi Company. The Montrose was then out of reach of the land stations, and the message was telegraphed from ship to ship until it reached the Montrose in mid-Atlantic.

Captain Kendall was then unable to reply direct to England, owing to the distance from land.

A similar message was sent to him on behalf of the "Daily Mail" by our Montreal representative, viâ the Belle Isle wireless station. In this message Captain Kendall was asked to transmit his reply through our Montreal representative, who is connected with the "Montreal Star," so that he might forward it on to London.

On Friday last Captain Kendall courteously sent by wireless to our correspondent at Montreal, viâ Belle Isle, his first telegram, explaining why he believed "Mr. Robinson" and his "son" to be Dr. Crippen and Miss Le Neve. His telegram, it will be remembered, concluded with these words: "This is the first account that has been transmitted from this ship to any newspaper."

The wireless telegram in question was published in the "Montreal Star" of Friday last, and also, by arrangement with the "Daily Mail," in the London "Evening News" on the same day, and this was the first and only account by Captain Kendall or anybody in the Montrose given to the world.

In the early hours of Saturday morning the first portion of another long remarkable wireless message from Captain Kendall was received by the "Daily Mail" through the same channel, and appeared in the Boulevard Edition. The wireless connection, however, with the Marconi station was lost in the middle of the message, and the remainder of the telegram was received on Saturday afternoon, thirteen hours later.

The full text of this remarkable exclusive message appears in another column of our issue to-day. The "Daily Mail" was the only newspaper which had a special correspondent on board the Laurentic, the vessel in which Chief Inspector Dew travelled to Rimouski. This correspondent sent a very interesting wireless message on Friday, telling the

Hawley Crippen's Homeopathy

Dr Hawley Crippen and Cora Turner were an unlikely couple. She was an aspiring singer and dancer, loud, large and blowsy. Crippen, by contrast, was small, well-spoken and unassuming.

However, 18 years after the pair married in 1892, Crippen stood accused of poisoning Cora and burying her dismembered body beneath the cellar floor. What had led to this extraordinary turn of events?

Crippen and Turner met in New York, married in Philadelphia and moved to London in 1897 so Crippen could pursue his career in the homeopathic mail order business. However, Crippen lost his job after too many absences from trying to launch Cora's stage career, for which she showed little aptitude. Crippen then took a job as a consulting doctor at a London clinic, where he met and fell in love with a demure 18-year-old assistant, Ethel "Le Neve" Neave.

Cora, meanwhile, busied herself socializing with the London theatre crowd, working as a fundraiser for the local Music Hall Ladies' Guild, and entertaining friends at the salubrious Crippen home at 39 Hilldrop Crescent, Holloway. Cora's affairs were common knowledge, but Crippen seemed unaware of their existence until he walked in on Cora in bed with one of their lodgers. From that point on, Cora and Crippen stayed under the same roof as man and wife in name only: Cora carried on with her dalliances, as Crippen did with Le Neve.

However, by 1910 life at Hilldrop Crescent had become unbearable. After Cora threatened to expose Crippen and Le Neve's affair, Crippen decided to take matters into his own hands. He acquired five grains of the poison hyoscine hydrobromide, a sedative, and set his plan into action. No one saw Cora alive again. A few days later the Music Hall Ladies' Guild received a letter allegedly from Cora, offering her resignation, as she was travelling back to America to tend to a sick relative. Meanwhile, Le Neve moved into Hilldrop Crescent and began behaving as Crippen's wife.

It was the appearance of Le Neve in Cora's jewellery and clothes that set off alarm bells at the Ladies' Guild. Something seemed amiss, and they badgered Crippen relentlessly for news about Cora, who seemed to have dropped off the edge of the world. That is, until Crippen sent the Guild a telegram saying that Cora had died suddenly in the United States. The guild was astonished by this news, especially as it said Cora would be cremated, which did not tally with her Catholic background. Sensing something was afoot, the Guild called the police at London's Scotland Yard.

When Scotland Yard Inspector Walter Dew visited Hilldrop Crescent, Crippen confessed he had told some lies. But these, he said, were to cover up the embarrassing fact that Cora had left him for a new life in America with one of her lovers. Crippen seemed believable, and Dew was willing to accept his story. However, when he returned to Hilldrop Crescent with some routine follow-up questions he found Crippen and Le Neve had fled the country.

OPPOSITE: The Daily Mail *newspaper dedicated multiple column inches to the Crippen arrest.*

This was enough for Dew to order for Hilldrop Crescent to be thoroughly searched. The results were grisly. All that remained of the body buried under the cellar floor was a torso, without a head, arms, legs or genitals, and with all of its bones removed. There was, however, an abdominal surgical scar that apparently matched a scar of Cora's. There were also traces of hyoscine in the body; this was the smoking gun that led directly to Crippen. A warrant went out for the couple's arrest.

Crippen, meanwhile, was aboard the liner SS *Montrose* sailing for Canada, he posing as a Mr Robinson and Le Neve as his teenaged son. Le Neve, however, made a less than believable boy; the couple's amorous embraces also indicated something other than a filial relationship. The captain of the *Montrose* was certainly not taken in; he used the telegraph, a brand-new technology, to wire Scotland Yard: "Have strong suspicions that Crippen London cellar murderer and accomplice are among saloon passengers. Moustache taken off growing beard. Accomplice dressed as boy. Manner and build undoubtedly a girl."

Dew was on the next fast boat for Canada and met Crippen and Le Neve as they docked. Dew said: "Good morning, Dr Crippen. Do you know me? I'm Chief Inspector Dew from Scotland Yard." Crippen replied: "I am not sorry; the anxiety has been too much."

Crippen's subsequent trial in London lasted four days and revealed what remains of Cora had been found: part of her liver and kidney, a curler with her

DID CRIPPEN DO IT?

In 2007, toxicologist John Trestrail decided to re-examine the evidence connecting Crippen to Cora's death. Crippen had always maintained his innocence, and certain aspects of the murder, including the mutilation of the body, did not fit the profile of other poisoners, who often used poison to avoid unnecessary butchery.

Forensic examiners from America's Michigan University compared DNA from the torso skin found in the cellar of Hilldrop Crescent with that of Cora's descendants. The results were conclusive: Cora's family's mitochondrial DNA, which remains unchanged through the generations, did not match that taken from the torso. Even more astoundingly, the team found that the DNA contained a Y chromosome, which meant the torso had belonged to a male. Many argue that the findings do not prove Crippen's innocence, but one thing is sure: he did not murder and dismember his wife and bury her torso under his cellar floor.

LEFT: *Cora 'Belle Elmore' Crippen was a theatrical socialite who had numerous affairs. She had been an odd match for Crippen.*

HYOSCINE

Hyoscine is also known as scopolamine and is derived from the deadly nightshade plant. In Crippen's day, hyoscine was used to treat motion sickness or prevent period pain, and given as a sedative to the violently insane. Hyoscine is effective as a poison only in high doses, such as the five grains Crippen ordered. At this dose, hyoscine would have caused a dry mouth, blurred vision, hallucinations, coma, paralysis and death. However, if a victim of hyoscine survives an overdose, they often make a full recovery. Those who do not believe Crippen murdered Cora point to hyoscine as evidence of his innocence: it is a strange choice of poison given the wide availability of the far more effective arsenic.

hair, and a fragment of Crippen's pyjamas. Crippen pleaded not guilty, but the jury was clearly swayed by the series of lies that he had fed the Music Hall Ladies' Guild. It took the jury under half an hour to find Crippen guilty; he was executed by hanging at Pentonville Prison on 23 November 1910.

Before he perished, Crippen did manage to convince the court that Ethel Le Neve had played no part in Cora's disappearance. After 12 minutes of deliberation, the jury acquitted Le Neve. She sailed to Toronto on the day of Crippen's execution and later returned to London under the name Ethel Harvey. She remarried and died in 1967. The Crippen house at 39 Hilldrop Crescent was destroyed by German bombs during World War II.

ABOVE: The arrested Crippen and La Neve are led off the Canadian SS Montrose. *The ship's captain had not bought Le Neve's disguise as a boy.*

Georgi Markov's Umbrella Death

On a September evening in 1978, Georgi Markov was waiting for a bus on Waterloo Bridge, London, when something sharp stabbed his thigh.

A thick-set man beside Markov dropped an umbrella, mumbled "sorry" in a foreign accent, and jumped into a taxi. Three days later, Markov was dead. What emerged was a chilling Cold War episode torn straight from the pages of an Ian Fleming spy story.

Georgi Markov was a Bulgarian dissident and writer who had been a major irritant to his country's communist government. The regime retaliated by censoring Markov's plays and halting publication of one his novels while it was still running off the printing press.

In 1969, Markov fled to exile. In London, he continued his criticism of his country's government during broadcasts on the BBC's World Service. He saved his most vitriolic comments for the inner circle of Bulgarian leader Todor Zhivkov, who pronounced Markov a "non-person" and sentenced him to six years in jail in absentia for his defection.

Markov was on his way home to his cottage in Dorset when the incident on Waterloo Bridge occurred. It emerged that Markov had been shot with an "umbrella gun" fitted with a cylinder of compressed air to deliver a 1.52-mm, pinhead-sized metal pellet into Markov's thigh. The pellet turned out to be a jeweller's watch-bearing that had had two tiny holes, each 0.34 mm in diameter, drilled into it to form an X-shaped well inside. Only a high-tech laser is capable of drilling such holes into a hard alloy.

The holes in the pellet were found to have been covered with wax, which would have melted inside Markov's leg from his body temperature. The wax gone, the pellet would have released its payload of around one-fifth of a milligram of plant toxin, later identified as ricin. Although the autopsy discovered no traces of ricin in Markov's body, his symptoms mirrored exactly what toxicologists would expect from the poison. Later, scientists injected a pig with a similar amount of ricin to prove their theory correct.

After six hours the pig had come down with a high fever, an elevated white blood cell count, and the same internal haemorrhaging that had killed Markov. Markov died only hours after being injected. The facts stacked up: ricin was well known to the CIA as a weapon developed by the Soviet Union and its allies. Moreover, the Bulgarian secret service had already attacked another dissident, Vladimir Kostov, by firing a similar pellet into his neck. However, unlike Markov, Kostov had lived to tell the tale. Despite the Markov case being re-opened in 2008 with the aid of the Bulgarian government, it has never been conclusively resolved.

OPPOSITE: Georgi Markov is pictured after his defection to Britain from Bulgaria. To date, nobody has been charged with his murder.

RIGHT: A KGB umbrella gun.

Ricin

Ricin is an extremely potent toxin extracted from the castor bean (Ricinus communis), which is used for castor oil. Ricin is most commonly manufactured into a white powder.

OVERVIEW:

The ricin found in castor beans is itself a powerfully toxic substance – just 20 beans can kill an adult. As a powder ricin is even deadlier, and a few grains weighing around 1.78 milligrams are usually fatal. Ricin is easily isolated and was considered as a coating for bullets and filling for bombs by the UK and US military during the early twentieth century. The toxin was weaponized by the Soviet Union's KGB and used by Communist countries during the Cold War. Assassination attempts were famously made on Bulgarian dissidents Georgi Markov (pages 130–131) and Vladimir Kostov during the 1970s. Ricin again emerged as a terrorist weapon early this century. Letters containing ricin were delivered to New York Mayor Michael Bloomberg and then US President Barack Obama in 2013; both were safely intercepted. Ricin packages were also detected in mail sent to the US Pentagon in 2018.

TOXIC EFFECTS:

Ricin poisoning usually occurs after ingestion or inhalation of the substance, or having it injected. Ricin has two poisonous elements: the first penetrates the body's cells, and then creates a tunnel down which the second toxin can travel. This toxin then stops the cell's ability to produce proteins, which in turn kills the cell. Once ricin enters the victim's bloodstream, its deadly effects spread throughout the body. However, ricin is a slow-burning poison and symptoms often do not present themselves for between four and 24 hours. A long, agonizing death over three to five days can follow, depending on the dosage.

SYMPTOMS:

The typical first symptoms of ricin poisoning include difficulty breathing, fever, coughing, tightness in the chest and nausea. Vomiting, bloody diarrhoea and low blood pressure can follow, and the victim's lungs sometimes fill with fluid. If the poison progresses it leads to respiratory failure and severe organ damage; the sufferer's spleen, liver and kidneys may stop working altogether, leading to death.

TREATMENT:

There is no antidote for ricin, and all medical professionals can do is flush the poison from the body. Charcoal to flush the stomach is sometimes administered.

FAMOUS POISONINGS:

- Soviet dissident and novelist Aleksandr Solzhenitsyn survived a KGB ricin attack in 1971. He later died of a heart attack in 2008.
- In 1981, CIA double agent Boris Korczak was shot in the kidney with a bullet containing ricin. This was almost certainly the work of the KGB, but Korczak survived because his body expelled the bullet as it would a kidney stone.
- In 2014, American teenager Nicholas Helman sent a scratch-and-sniff birthday card contaminated with ricin to his ex-girlfriend's new boyfriend. After being discovered, Helman was arrested following a shoot-out with police.

OPPOSITE: An illustration of the castor bean plant.

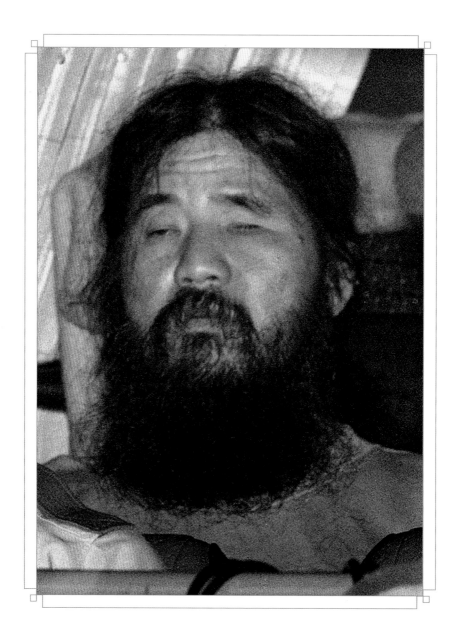

The Tokyo Sarin Poisonings

In March 1995, the half-blind guru Shoko Asahara received an alarming tip-off. His disciple working in the government had learned of a police raid on the headquarters of Asahara's Aum Shinrikyo doomsday cult.

In response, Asahara ordered five disciples to commit an act of mass terrorism that he hoped would bring about a global apocalypse. The target would be the Tokyo subway system; the poison, the deadly nerve gas sarin.

The morning rush hour of 20 March seemed like any other to the Tokyo commuters crammed onto the world's busiest subway. But riding with them were five Aum Shinrikyo members, each carrying newspaper-wrapped plastic bags containing sarin. Sarin is a deadly, volatile poison developed by the Nazis; once the liquid is released it evaporates into a gas that quickly spreads over a wide area. Thousands of commuters were now at serious risk of death.

The five Aum members did not fit the typical profile of suicidal terrorists: three had degrees in physics; another was a senior doctor. Nor did the terrorists plan to end their own lives in the subway attacks. Each one carried a syringe filled with atropine sulphate, the antidote to sarin, and had a driver waiting above ground to whisk them back to Aum headquarters. Here, they would be greeted as heroes by Shoko Asahara: the charismatic founder of Aum Shinrikyo, an officially recognized religion with more than 40,000 followers worldwide.

At 8 am, the attack was put into action. Each terrorist was travelling on a different underground line to cause the maximum possible impact; at his feet were two or three packages of sarin. These were now punctured with umbrellas that had sharpened tips. As the liquid spilled out and vaporized into lethal, invisible clouds of gas, the terrorists fled to safety. However, one had been so clumsy in piercing the package that he had already poisoned himself and would need to administer the antidote in his getaway car.

The passengers left on the trains quickly began to feel irritation of their eyes and noses; soon they were shivering and coughing violently; some began to vomit. Many later reported the air becoming thick and smelling like paint thinner. Panic quickly set in, with passengers screaming at each other to get off the trains.

On some of the trains the emergency stop button was pushed and passengers piled out and up towards fresh air. Many collapsed onto the station platforms, gasping for breath with blood streaming from their noses. Soon there were piles of bodies in the carriages and on platforms; many victims were convulsing violently, and others were unconscious. Meanwhile, some trains continued on towards the city centre with the mysterious puddles of sarin not yet understood to be the cause of the "travelling gas chambers", as they were later dubbed by the country's press.

By unleashing the sarin on the Tokyo subway, Shoko Asahara hoped to kill thousands of people and then implicate the United States military in the attack. This, he reasoned, would not only deflect attention away from his own nefarious plans, but

OPPOSITE: *Aum Shinrikyo leader Shoko Asahara is photographed after his arrest in 1995. He was executed by hanging in 2018.*

cause a nuclear war between the US and Japan. Asahara promised his acolytes that they would survive the Armageddon that would follow, and live with him in the mythical kingdom of Shambala.

However, Asahara would not reach his desired death toll on the Tokyo subway. Although this was the deadliest terrorist attack on Japanese soil, only 13 people were killed, with a further 5,500 injured. Many of the injured suffered from long-term illness as a result, especially eyesight problems and post-traumatic stress disorder. The reason for the low number of fatalities was that the Aum sarin had not been processed properly and was of a low grade. This was perhaps surprising; Asahara had launched a previous attack in 1994 using high-grade sarin in an attempt to assassinate some judges presiding over an Aum legal dispute over land. Then, the sarin was despatched into the air around the judges' homes in Matsumoto via fans fitted to trucks carrying the gas. Eight people died and 500 were injured during this attack.

The Matsumoto affair also led police straight to Asahara following the Tokyo attack. In May 1995, thousands of police officers launched raids on Aum Shinrikyo's buildings, including its headquarters in Tokyo. Inside, police found explosives, millions of dollars in cash, a Russian military helicopter, and chemical weapons, including enough sarin to kill over four million people. Also being manufactured in the Aum laboratories were the drugs LSD, methamphetamine and a type of truth serum. Cells containing prisoners were also discovered.

More than 150 Aum members were arrested during the raids. Shoko Asahara himself was finally discovered hiding in a small space behind a fake wall, along with silk pyjamas, a sleeping bag and a pile of cash. During his trial, Asahara mumbled incoherently and his defence appealed his death sentence on the grounds that he was mentally unfit. However, as Asahara had been able to communicate with staff at the detention centre, the verdict was upheld. Shoko Asahara was executed by hanging on 6 July 2018, along with six other members of the cult.

SHOKO ASAHARA

Shoko Asahara, born Chizuo Matsumoto in 1955, founded the doomsday cult Aum Shinrikyo in 1987. Registered as a religion in 1989, Aum mixed elements of Hinduism and Buddhism with apocalyptic Christian prophesies. Asahara wrote several religious books claiming that he was the new Christ who would take on the sins of his followers and transfer spiritual power back to them. Followers were invited to ingest large quantities of LSD during initiation ceremonies, which sometimes included being hung upside down and given shock therapy. Afterwards, followers were encouraged to wear specially designed electrode caps that would re-programme their minds via a constant electrical charge.

Aum followers were convinced to leave their families and turn their savings over to Asahara; they could also buy vials of his bathwater as a further aid to enlightenment. Asahara typically attracted young, idealistic people dissatisfied with what they saw as a materialistic, conformist society. Sometimes educated to PhD level, Aum members would help synthesize drugs and poisons for Asahara and also oversee work at a computer assembly plant. Aum's income from the plant and various other businesses amounted to millions of dollars every year. It was with this money that Asahara began preparing for a third world war, which he said would be started by the United States.

From 1993, Aum began its manufacture of chemical weapons, including anthrax, VX gas and sarin. After the 1995 attack, Aum rebranded itself as Aleph; it is still active in Japan. Aum Shinrikyo continued in Russia until 2016, when authorities raided its compounds and declared it to be a terrorist organization.

OPPOSITE: Japanese defence personnel begin the clean-up of the Tokyo subway system in 1995.

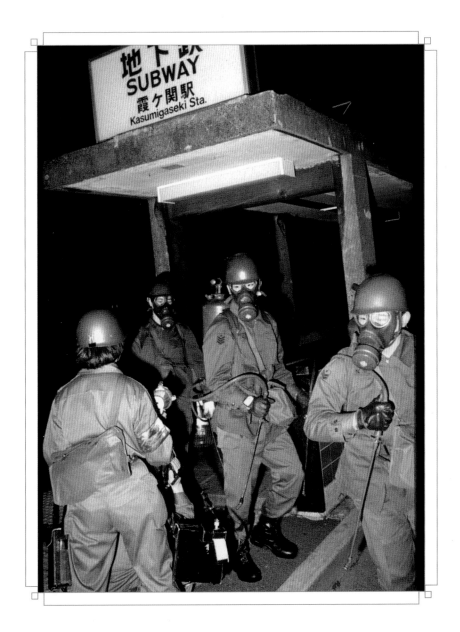

Sarin

Sarin is a clear, colourless man-made nerve agent developed for chemical warfare. It is named after the initials of the scientists who discovered it: Schrader, Ambros, Rudiger and van der Linde.

OVERVIEW:
Sarin was first developed by Nazi scientists, although it was still in its infant stages by the end of the war and was never deployed in the field. In the 1950s, the United States, the USSR and NATO adopted sarin as a chemical weapon, before the stockpiling of the poison was outlawed in 1997 under the Chemical Weapons Convention. Over 14,000 tonnes of sarin has since been destroyed worldwide. However, this was not before sarin was used against ethnic Kurds by Iraq in 1988 (see Famous Poisonings, below). Sarin as a poison can be swallowed, absorbed through the skin, or released as a liquid, which then immediately vaporizes into an invisible, spreading gas. Exposure to sarin is extremely dangerous; even at low concentrations it can kill an adult in under 10 minutes. As a gas, sarin is 26 times more deadly than cyanide and is widely considered to be a weapon of mass destruction. Sarin made a re-appearance in warfare when it was used during the Syrian Civil War by government forces. Incidents in 2013, 2017 and 2018 have been reported.

TOXIC EFFECTS:
Sarin works by blocking the enzyme (acetylcholinesterase) that naturally destroys a neurotransmitter called acetylcholine. This means that the neurotransmitter, which normally delivers messages between nerve cells, keeps repeating its message over and over again. For example, if the message is "release a little saliva to moisten the mouth", the repeated message becomes "release saliva in perpetuity". Although this sounds relatively harmless, acetylcholine build-up puts human muscles and secretions into overdrive: the eyes, nose and mouth run and the bowel and bladder evacuate themselves involuntarily.

SYMPTOMS:
Victims of sarin poisoning quickly experience pin-point pupils, watery eyes, blurred vision, a running nose and drooling mouth, chest pain and rapid breathing, coughing and vomiting, diarrhoea and uncontrolled urination, confusion, weakness, headaches and altered heart rate, followed by convulsions, paralysis, respiratory failure and death.

TREATMENT:
Those poisoned by sarin should try and leave the contaminated area as quickly as possible, especially to find higher ground, as sarin sinks when it is in gas form. Clothes should be removed, eyes flushed with water and skin washed with soap. Atropine sulphate is an antidote to sarin poisoning as it liberates acetylcholinesterase and allows the enzyme to work as normal.

FAMOUS POISONINGS:
- In March 1988 Iraqi forces under Saddam Hussein attacked the Kurd-held Iraqi town of Halabja with nerve gases that included sarin. More than 5,000 people were killed in this "Bloody Friday" attack, and countless others suffered from blindness and long-term illness.
- In 2013 during the Syrian civil war, a chemical weapons attack was launched on rebel forces in Ghouta, outside the capital Damascus. Rockets found to contain sarin were used in the attack, which killed between 280 and 1,729 people.

OPPOSITE: A artillery rocket from the Syrian conflict. A gas canister has replaced the usual explosive warhead.

Harold Shipman's Sentence

The death of 81-year-old Kathleen Grundy came as a surprise to her daughter, Angela Woodruff. Grundy had been in good health when she died suddenly at her home, attended by doctor Harold Shipman.

A few weeks later, a badly-typed will emerged awarding Grundy's entire estate to Shipman. Alarmed, Angela called the police. Mrs Grundy's body was exhumed and revealed high levels of diamorphine. Kathleen had been murdered.

When police charged Shipman with murder after Grundy's autopsy, they did not realize that they were apprehending one of the most prolific serial killers in history. Shipman was a respected doctor whose solo practice in Hyde, Manchester was especially popular with elderly patients. The elderly also made up the majority of his 218 known victims; some people estimate that he murdered over 250 patients between 1974 and 1998.

Born in 1946, in Nottingham, Harold Frederick Shipman was a lorry driver's son who developed a fascination with drugs after seeing his dying mother injected with morphine during her struggle with cancer. Shipman studied medicine and began practising as a GP in 1974. Soon afterwards he was caught writing himself a prescription for Demerol, an analgesic drug he was abusing. Shipman attended drug rehabilitation and in 1977 became a GP in Hyde, Manchester.

Shipman was said always to have time for elderly patients; he would even visit them at home. However, in 1998 a local funeral director noticed that many of Shipman's elderly female patients appeared to be dying; most were found clothed and upright in an armchair at home at their time of death. Shipman's death certificate often stated "old age" as the cause of death. In fact, the injections of morphine and diamorphine administered by Shipman would have caused respiratory depression in his victims: simply, they stopped breathing.

An inquiry was launched by police in 1998, but they found no wrongdoing. Then, on 24 June, Kathleen Grundy died and Shipman was exposed. A search of Shipman's house turned up the typewriter that he had used to forge Grundy's will; Grundy's patient records had also been retroactively altered to indicate that she was a morphine addict. A large stockpile of diamorphine was found at Shipman's home as well as £10,000 worth of jewellery, which did not belong to his wife.

Was money the motive for Shipman's murders? Apart from Grundy and the stolen jewellery there is no evidence for this. Psychologists variously suggested that Shipman was avenging his dead mother, or euthanizing the elderly so they wouldn't be a burden on the National Health Service. Others speculated that he simply enjoyed playing God. His true motivation, however, will never be known. After receiving a life sentence for the murders, Shipman hanged himself in his cell in 2004.

OPPOSITE: Doctor Harold Shipman was Britain's most prolific serial killer.

Threatening George Trepal

Not every story of murder through poisoning involves political intrigue, mass slaughter or a clandestine spy operation. Sometimes, the motivation stems from something as banal as a dispute between neighbours.

This in fact lay at the heart of a mysterious poisoning that occurred in Florida, USA in 1988. It was a crime that had investigators stumped, until attention turned to George Trepal: a neighbour with a grudge.

George Trepal began his vendetta with his neighbours the Carrs after their dogs chased his cats and their teenagers played loud music. After several heated confrontations, the Carrs received a typed note on their doorstep: "You and all your so-called family have two weeks to move out of Florida forever or else you will all die. This is no joke."

The Carrs did not take the note seriously, but a few weeks later 41-year-old Peggy Carr began experiencing a strange stomach upset, numbness in her hands and feet, and a terrible pain in her legs.

At first the doctors told Peggy her symptoms were psychosomatic, but they began to take her seriously when her hair fell out. Soon she was unable to speak and was hospitalized. Then Carr's two teenagers also started having the same symptoms. The test revealed that all three family members had high levels of the poison thallium in their blood.

Suspecting foul play, investigators traced the poison to an eight-pack of bottled Coca-Cola. Oddly, thallium was found in unopened bottles with their caps sealed as well as those currently being consumed. A few months later, Peggy Carr died from thallium poisoning. This was now a murder case. Investigators homed in on Trepal after he suggested that someone might poison a neighbour to "make them move out". It also emerged that Trepal had served three years for working in an illegal amphetamine laboratory in the 1970s. The motive and expertise seemed obvious, but the lack of evidence prevented an arrest.

To investigate Trepal further, an undercover officer befriended him at a murder mystery weekend he was organizing at a Holiday Inn. Among the clues used for the game was the disturbing note: "When a death threat appears on the doorstep, prudent people throw out all their food … Most items on the doorstep are just a neighbor's way of saying, 'I don't like you. Move or else.'"

The undercover officer was later able to rent the Trepal home after they moved out. Here, she discovered the smoking guns: a bottle containing thallium and a machine for capping bottles. The evidence was circumstantial, but it was enough to sentence George Trepal in 1991 to be executed for first-degree murder of Peggy Carr. He remains on death row at the time of writing.

ABOVE: *The Coca-Cola bottles tainted with thallium by George Trepal.*

OPPOSITE: *Homicide investigator Sergeant Susan Goreck went undercover to catch Trepal.*

Teenage Japanese Poisoner

The English serial killer Graham Young seems an unlikely idol for a Japanese teenager. But in 2005, a 16-year-old girl from Shizuoka prefecture began copying Young's crimes.

She laced her mother's food and drink with thallium, Young's poison of choice. As her mother's health deteriorated, the teenager recorded the results of the poisoning in a blog. This was also how she was caught.

The Shizuoka teenager, whose identity as a minor is secret, developed an obsession with Graham Young after reading his biography. Using thallium and sometimes antimony and belladonna, Young murdered three people in the 1960s and 1970s and poisoned dozens of others.

To emulate her hero, the Shizuoka teenager purchased 50 mg of thallium from a local pharmacist, explaining that it was for a school science project. "The man in the pharmacy wasn't aware he had sold me such a powerful drug," she wrote in her blog.

There was nothing to suggest that the teenager would contemplate using the poison on her mother. A member of an elite Japanese high school, the girl was described as hard-working, with a bent for science; it was assumed that she would become a chemist.

As well as science, the teenager also enjoyed dissecting animals. "To kill a living creature. The moment of sticking a knife into something. The little sigh. I find it comforting," she wrote. A severed cat's head was later found in her room by police.

The teenager began writing about the effects of the thallium on her mother shortly after she began surreptitiously feeding it to her. "My mother has been sick since yesterday. She has a rash all over her body," the teenager wrote on 19 August. On 12 September, she updated her blog: "My mother has been complaining her legs are no good for two or three days. It is almost impossible for her to move."

A few days later, the mother became severely unwell and was admitted to intensive care. "Mother seems to have started hallucinating", the teenager noted in her blog. While in hospital, the teenager continued to feed her mother thallium, while also administering it to herself to place herself beyond suspicion. However, the teenager's brother did not buy it. "I took a photo of her [the mother] today, as I did yesterday. My brother said I had a penetrating stare and that he was horrified."

The court case was open and shut. The teenager was the only one to buy thallium in Shizuoka for five years, and had left her name and address at the pharmacy. In 2006, she was sent to a correctional facility for treatment where she was reported to spend her time "writing diary-like notes and poems".

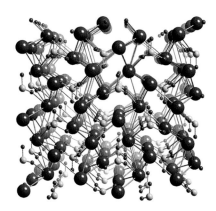

ABOVE: A magnified illustration of a thallium crystal.

OPPOSITE: English serial killer Graham Young, who was nicknamed the 'Teacup Poisoner'.

Thallium

Thallium is a soft, bluish-white metal that bears the chemical symbol TI and is found in trace amounts in the Earth's crust.

OVERVIEW:
Discovered in 1861, thallium received its name from *thallos*, or "green shoot" in Greek, after the bright green colour that the metal produces when burning. In the past thallium has been obtained as a by-product of smelting other metals. However, burning or melting thallium is an extremely dangerous process; the metal is highly toxic and can easily be absorbed through the lungs and skin as well as being ingested. Thallium has been used in pesticides and rat poison and in certain electronic devices in the semiconductor industry. It is primarily released into the atmosphere through smelting and the burning of coal, as well as through food. Plants easily absorb thallium, and once released into the air, water and soil the toxin takes a long time to break down. It is known to build up in fish and shellfish. Because it is odourless, colourless and reasonably difficult to detect, thallium was once considered a highly effective poison, called the "poisoner's poison".

TOXIC EFFECTS:
Thallium breaks down the cells in the human body, especially hair follicle cells and those in the central nervous system. Taking up to three weeks to have an effect, thallium poisoning is usually shown through extreme hair loss, followed by respiratory depression, kidney, liver and heart failure and death. Thallium can also cause brain damage, personality disorder and even psychotic behaviour in some victims.

SYMPTOMS:
Symptoms of thallium poisoning begin with hair loss, followed by nausea, abdominal pain, vomiting, bloody diarrhoea, spasms, convulsions, pneumonia and death.

TREATMENT:
The drug Prussian blue is an antidote for thallium poisoning and is given alongside stomach pumping to clean the gastrointestinal tract.

FAMOUS POISONINGS:
- In Australia in the 1950s, there were a number of recorded instances of attempted murder by thallium poisoning. Several of these "Thallium craze" cases featured suburban Australian women poisoning abusive family members.
- In 1971, anti-colonialist Cameroonian leader Félix-Roland Moumié was assassinated using thallium by a former French secret agent in Geneva. It was suspected but never proven that his murder was at the request of the Cameroonian government.
- Saddam Hussein poisoned Iraqi dissidents with thallium before expelling them from the country and leaving them to die weeks later at their point of destination.
- In 2004, Russian soldiers became ill after finding a tin of mysterious white powder in a rubbish dump near their military barracks. The soldiers used the thallium as talcum powder on their feet and added it to their cigarette tobacco, but all miraculously survived the experience.

OPPOSITE: A piece of thallium in a test tube bearing the TI symbol.

Chapter 6: Twenty-first Century Poisons

Poisons of the twenty-first century have often been linked with the great threat of the modern age: terrorism. In 2001, the 9/11 Al-Qaeda attacks on America were quickly followed by an altogether more pernicious and potentially even more dangerous threat to the country's security: anthrax.

First appearing at the offices of American news organizations, letters laced with anthrax were assumed to be the next wave of attack waged by terrorist jihadists. Anthrax is a disease caused by the weaponized spores of the *Bacillus anthracis* bacterium. Once ingested, the bacteria grow and multiply in the blood vessels, and the body shuts down; the victim drowns in their own body fluids. However, investigators concluded that the anthrax was not developed in the laboratories of an Islamic fundamentalist but instead those of the US military.

Such is the double dilemma of terrorist poisoning: first the correct poison must be identified, and then the correct culprit. In 2004, Ukrainian presidential candidate Viktor Yushchenko was terribly disfigured after swallowing dioxin-laced rice. The perpetrators were widely believed to be from the highest echelons of the Russian government, but the Kremlin claimed that Yushchenko had probably just eaten a bad batch of sushi.

Russia was also forced to deny wrongdoing after two separate terrorist attacks were launched on British soil. The first was former Russian spy Alexander Litvinenko, poisoned with polonium in 2006; the second was double-agent Sergei Skripal and his daughter Yulia, poisoned with novichok in 2018. In both cases the military-grade poisons used could only have been developed in a laboratory with government-level resources. Both poisons were also known to have been manufactured in the Soviet Union.

No-one has yet proved that the Russian government carried out these crimes, but experts say only the most senior members of the Kremlin would have authority to approve the release of such top-secret toxins. Before the 2018 Skripal poisoning, few people had even heard of novichok: now it is a household name. Samples of novichok have been obtained and developed in laboratories across NATO countries, and scientists now know how to test for it.

This is the role of poison as a murder weapon in the twenty-first century: new substances are developed, administered and then examined so the toxicology can be updated. Our early ancestors added poison to the tips of spears and arrow points. Now, poison has become a government resource, developed in top-secret military facilities and smuggled across international borders to be used on enemies of the state. The threat of a vast city-wide attack using military-grade poison has yet to be realized; however, it remains a real and dreadful possibility, a dark new chapter in the long history of poisoning.

OPPOSITE: Hazardous material workers in protective suits respond to the 2001 anthrax attacks in Washington, DC.

The Washington Anthrax Attacks

Only one week after the terrorist attacks of 11 September 2001, letters containing deadly anthrax spores were mailed to the offices of US news agencies and senators.

Posted from a non-existent "Greendale School", the handwritten letters declared "death to America" and bore the hallmarks of a jihadist extremist. The country reeled under the news of this terrible new terrorist threat. The United States was once again under attack.

On 2 October, Bob Stevens, editor of Florida tabloid the *Sun*, grew short of breath and started vomiting. A few days later he became the first person to die of anthrax poisoning in the US in 25 years. Soon others were hospitalized, including postal workers. It later emerged that some of the anthrax letters had been punctured by mail-sorting machines. Anthrax contamination now spread across the wider US mail service.

Meanwhile, letters containing weapons-grade anthrax arrived at the offices of the *New York Times* and Senator Tom Daschle. Some of the anthrax came as a clump of brown granules; in others it was a white powder resembling talcum powder. The latter was highly potent anthrax that would vaporize once the letter was opened and could be easily inhaled by its victim. Once inhaled, the anthrax would create lesions in the victim's lungs and brain and quickly devour the body from within.

On 21 October, postal worker Thomas Morris died from anthrax inhalation. The outbreak would claim 16 further lives, including a Connecticut woman whose mail apparently crossed paths with an anthrax letter, and a hospital clerk in Manhattan. Hoax letters began appearing nationwide; the police became swamped with calls about suspicious mail.

Under enormous pressure to locate the culprit, FBI investigators surmised that the terrorist was most likely to be a US citizen with links to the scientific community. The letters had been addressed to the senators in the same style all US school-children would know; oddly, some letters suggested that the recipient immediately take penicillin after reading. Probably only American scientists had a working knowledge of weapons-grade anthrax. The gaze of the authorities fell on doctor Steven Hatfill, a one-time employee of the military's elite Medical Research Institute of Infectious Diseases (USAMRIID), which also stockpiled anthrax.

After being publicly vilified for several years, Hatfill was cleared of suspicion of the anthrax poisonings. It wasn't until 2008 that a new suspect emerged: bio-defence researcher Bruce Ivins. Ivins had worked with anthrax and appeared to fit the profile; however, he committed suicide before he could be brought to trial. Federal prosecutors closed the file, but critics argued that Ivins did not have the equipment or know-how to coat the powdered anthrax with the silicon that would make it vaporize when the letters were opened. Many believe that the true culprit is still at large.

OPPOSITE: The letter that was sent to US senators Tom Daschle and Patrick Leahy in 2001, along with powdered anthrax.

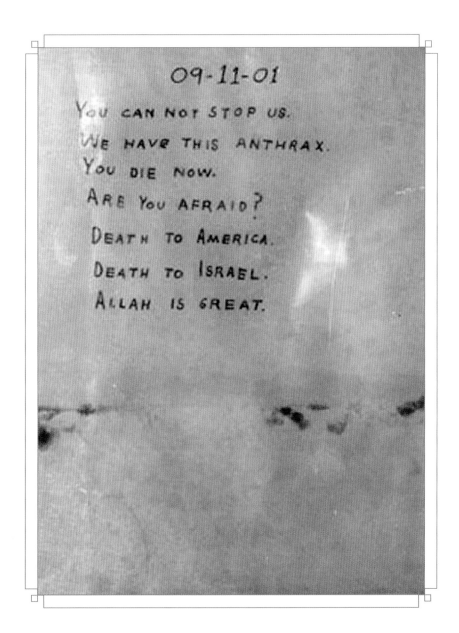

Anthrax

Anthrax is a disease caused by the bacterium Bacillus anthracis. *The bacterium can produce lethal spores, which retain their potency for many years. These spores can be weaponized into a powder or liquid and used as a poison.*

OVERVIEW:

Anthrax spores occur naturally in the soil, and most commonly affect grazing animals such as sheep and cattle as an intestinal infection. Humans traditionally contract anthrax from these animals by eating the meat or handling the wool, hair or carcasses. Anthrax is an ancient disease mentioned in the Bible and by ancient Greek and Roman authors. The bacterium was identified in 1863 and isolated as an organism in 1876. A vaccine against anthrax was developed by microbiologist Louis Pasteur in 1881. Anthrax has been weaponized by a number of countries, and the United States and the Soviet Union both had significant anthrax weapons programmes during the Cold War. The toxin can be easily manufactured, and its spores are small enough to spray in aerosol form over a large area; the fatal effects of a large attack would be on par with those of a nuclear explosion. Anthrax can also be delivered through guided missiles, bombs, crop dusters or similar aircraft. Despite attempts by various terrorists, a large-scale anthrax attack is yet to succeed.

TOXIC EFFECTS:

The three most common types of anthrax include: cutaneous, or skin, anthrax; gastrointestinal anthrax; and inhalation anthrax. The first two forms are the most common and are treatable; the latter is extremely dangerous as it involves the ingestion of multiple anthrax spores. Once in the bloodstream, anthrax devours the body from within as the bacilli propagate and become trillions of parasitic microbes in the blood. At the moment of death these appear as large, writhing worms growing, multiplying and taking over the victim's blood vessels. The initial flu-like symptoms lead to an agonizing death as lesions appear on the lungs and brain, cells in the body's immune system explode, septic shock emerges and the victim literally drowns in their own fluids.

SYMPTOMS:

Inhalation anthrax first presents itself with flu-like symptoms including a sore throat, muscle aches and fatigue. These develop into a shortness of breath, vomiting, skin sores, delirium, fever, shock and death.

TREATMENT:

The anthrax vaccine adsorbed (AVA) is the vaccine developed to protect United States military personnel. It is not widely available to the public. Anthrax also responds to Cipro, an antibiotic that protects against infection caused by inhaled anthrax.

FAMOUS POISONINGS:

- Anthrax was mentioned in the Biblical book of Exodus, and many scholars believe that the fifth of the "10 plagues of Egypt" may have been an outbreak of anthrax.
- In 2006, a New York drum-maker became sick with anthrax after processing animal hides to turn into drum skins. He survived the infection.
- In 2009 and 2010, an outbreak of anthrax poisoning occurred among intravenous drug users in the United Kingdom and Germany. Although they appeared to have cutaneous anthrax, they did not show the tell-tale raised sore with a black centre. Doctors now believe that the anthrax spores were in the heroin they injected.

OPPOSITE: The lethal anthrax spores seen under a microscope.

Iceman Richard Kuklinski

Richard Kuklinski's fatal mistake was not letting one of his victims' bodies thaw completely before dumping it.

Kuklinski, a Mafia hitman who may have murdered over 200 people, favoured the freezer method so he didn't have to deal immediately with body disposal. He became known as "The Iceman". But shards of ice left in a victim's not-quite-thawed body led to his capture.

Richard Kuklinski killed his first victim in 1948 when he was 13 years old. This was the teenaged leader of a rival New Jersey gang, whom Kuklinski bludgeoned to death with a wooden clothes hanger before cutting off his fingers and pulling out his teeth so the body could not be identified.

Kuklinski had been brought up by an abusive, alcoholic father and thought little of physical violence. As a boy, he routinely tortured cats by tying their tails together and throwing them over a clothes-line so they would tear each other to shreds. As he grew into a 6 ft 6 in heavy-set adult, Kuklinski became a popular hitman for the Mafia crime families of New York, including the DeCavalcante family, who provided the inspiration for the *Sopranos* TV series.

Kuklinski preferred the use of poison to avoid the mess of more physical murder methods. According to his own testimony, these included using guns, explosives, tyre irons, asphyxiation, hand grenades, and a bomb tied to a remote-controlled toy car, although on occasion he simply beat people to death with his bare hands "for the exercise". He also tied up victims and fed them to cave rats, which, he said, would completely dispose of a human body in under two days.

Cyanide was Kuklinski's poison of choice, because it was easy to administer and difficult to reveal in toxicology results. Kuklinski would sometimes spike a victim's hamburger with cyanide or conceal it in a nasal spray bottle, which he would then squirt directly into a victim's face. "You spray it on someone's face and they go to sleep," he later explained.

Kuklinski had learned about cyanide from fellow hitman "Mister Softee", so-called because he drove around in an ice cream van for cover. Mister Softee also taught Kuklinski about leaving the bodies of victims in industrial freezers: if left frozen for months, the body would not reveal the time of the murder. However, in 1986 after 38 years of killing, Kuklinski became careless. He did not thaw out the body of a known associate he had killed, and ice crystals were found in the cadaver's throat. The trail led to Kuklinski and then his freezers. Kuklinski was jailed for 18 years for six murders before dying in prison. He granted every interview while inside, to add to the extraordinary legend of "The Iceman".

ABOVE: Richard "Iceman" Kuklinski is taken into custody in 1986.

OPPOSITE: *New York's* Daily News *reports Kuklinski's arrest.*

MANHATTAN ★ ★ ★ SPORTS FINAL

COLOMBIA'S REIGN OF TERROR:
COCAINE KILLS
PART 4 OF WHY THE SMUGGLERS ARE WINNING — Starts on Page 7

DAILY ○ NEWS

35¢   NEW YORK'S PICTURE NEWSPAPER®   Thursday, December 18, 1986

Nancy: They deceived Ron
Page 3

BURGER MURDER

N.J. man held in killings of 5 with gun & cyanide

Story on page 2

HASENFUS IS FREE

Nicaraguan President Daniel Ortega as he handed over gunrunner Eugene Hasenfus (left) to Sen. Christopher Dodd (right) in Managua yesterday. At far right is prisoner's wife, Sally. Hasenfus will arrive home today and may be summoned before congressional committees investigating the

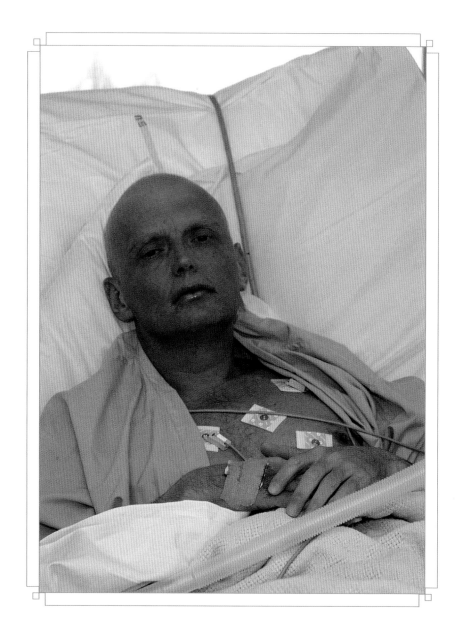

Alexander Litvinenko Assassination

When Alexander Litvinenko met two colleagues for a cup of tea at London's Millennium Hotel, he could not have imagined that it was to attend his own assassination.

For the tea had been spiked with a rare radioactive poison called polonium-210. During the days it took for Litvinenko to die, a real-life spy story emerged of murder and corruption that reached the highest levels of the Russian government.

Former spy Alexander Litvinenko had had a target on his head since he fled from Russia in 2000; his ex-colleagues used to fire at a picture of him at their Moscow firing range. There, Litvinenko was known as a traitor of the highest order; the former officer had defected to England to become a writer, a British MI6 agent, and a relentless critic of Russian president Vladimir Putin.

Litvinenko was a valuable asset to MI6: he had former colleagues from Russian intelligence services who were willing to sell secrets for cash. He also had an insider's knowledge of the Kremlin's links to organized crime. Most recently, Litvinenko had provided government spies in both the UK and Spain with information about the Russian mafia's operations in Spain. The trail was thought to lead to Putin himself, and Litvinenko had offered to testify before a Spanish prosecutor in late 2006. But not if he could be silenced first.

ABOVE: *The atomic structure of polonium-210.*

OPPOSITE: *Alexander Litvinenko during his final days in hospital.*

Russian assassins Andrei Lugovoi and Dmitry Kovtun arrived in London on 28 October 2006, carrying what Kovtun had told a friend was "a very expensive poison". Lugovoi and Litvinenko had worked together for the oligarch Boris Berezovsky; now Lugovoi had set up a meeting at the Millennium Hotel with his former colleague to discuss foreign investment opportunities in Russia.

Lugovoi arrived at the Millennium Hotel at 3.32 pm on 1 November 2006, and Kovtun 15 minutes later. Both visited the men's toilet in the lobby, where massive radiation contamination was later found in the cubicles and hairdryer. The men then ordered drinks, including a pot of green tea, and sat down in the hotel's Pine Bar.

Litvinenko arrived at 3.59 pm; Lugovoi met him and led him to their seats. The meeting lasted for 20 minutes. Litvinenko, who was short of money, declined to order anything. Lugovoi then told Litvinenko that there was tea on the table, and he was welcome to drink some. Lugovoi asked for a clean cup; a waiter brought one and poured the tea. Litvinenko then drank three or four sips of the cold, bitter-tasting green tea and put the unfinished cup down. Litvinenko later told police he had a feeling that the men wanted to kill him. He was right.

In the days after the meeting at the Millennium Hotel, Litvinenko became gravely ill. He was admitted to Barnet Hospital and then University College

Hospital for further investigation. Litvinenko's symptoms indicated thallium poisoning, but there was no evidence of this. Doctors were stumped. Meanwhile, Litvinenko gave a full and frank account of his meeting and poisoning, his former life as a spy, and later, reluctantly, his current work with MI6. Transcripts of the police interviews were released eight years later. In them, Litvinenko holds Vladimir Putin directly responsible. He also had to halt the interviews for attacks of diarrhoea. By 20 November Litvinenko became so ill his heart rate had become permanently irregular, his bone marrow was under severe attack, and his major organs were failing. Cancer doctors brought in to diagnose Litvinenko noted that he had the appearance of a victim of late-stage cancer: he was a deathly grey, his hair and eyebrows had fallen out and his skin was pockmarked. But still no-one could identify the poison that was killing him.

At this time, Litvinenko's blood and urine was sent to England's Atomic Weapons Establishment to be tested with spectroscopy. Initially no radioactive gamma rays were detected, but by chance, a scientist who had worked on Britain's atomic bomb programme recognized a barely discernible gamma ray spike in the spectroscopy reading. He identified this tiny spike as polonium-210, which was used in the creation of early nuclear weapons.

For Litvinenko, however, the diagnosis made little difference. On 22 November, he was struggling for consciousness and said goodbye to his wife Marina. Around midnight he had two heart attacks and was put into an induced coma. On the evening of 23 November after a third and final heart attack, Litvinenko died.

To date, and despite a British investigation that has implicated Putin and the Russian government in Litvinenko's murder, the Kremlin has denied any involvement in the crime. However, the polonium-210 used in the murder was traced to a nuclear reactor in the Russian town of Sarov. Only the highest-ranking officials would have the authority to order the release and deployment of this reactor's polonium. Litvinenko, a hated enemy of the state, became the first man in history to be assassinated with it.

THE POLONIUM TRAIL

The choice of polonium in Litvinenko's poisoning was described as "work of genius". The poison could be carried in a small bottle of water, is extremely hard to trace in the human body, and takes some time to take effect after being administered. However, the assassins hired to carry out the execution displayed levels of stupidity that bordered on incompetence. Andrei Lugovoi needed four attempts to successfully deliver the polonium to Litvinenko, flying to London each time to have a meeting with him. On his last attempt, Lugovoi succeeded. However, despite boasting that he was carrying an expensive secret weapon, Lugovoi appeared to have no understanding of its great ability to contaminate anything with which it came into contact. While pouring the polonium into Litvinenko's teapot, Lugovoi accidentally contaminated himself and left a trail of radiation that stretched across London. Everything Lugovoi touched immediately became contaminated with polonium: his credit cards, airline seats, restaurant tables, hotel light switches, and even people he had shaken hands with. Once police knew what they were dealing with, they were able to follow this radioactive glow to reconstruct an exact map of Lugovoi and Kovtun's movements, not just on their 1 November visit but on previous visits as well. Polonium from an a visit on 17 October was found on a shisha pipe at a Soho bar and later on the tables, doors and chairs of a late-night club. Large amounts of polonium were then tipped down the sink of the assassins' adjoining bathrooms at the Great Western Hotel. Perhaps, realizing the toxic implications of this action, the pair moved suddenly to another hotel the next day.

OPPOSITE: Litvinenko with copies of Blowing Up Russia: Terror from Within. *The book alleged Vladimir Putin orchestrated the 1999 Russian Apartment Bombings to justify the Second Chechen War and support his rise to power.*

Polonium-210

Polonium-210 is one of the radioactive isotopes of polonium, a silvery-grey metallic element that carries the chemical symbol Po. Polonium is extracted from uranium ore.

OVERVIEW:

Polonium-210 is a rare, highly radioactive substance that omits tiny, positively charged particles. It is found in very small doses in the soil and atmosphere but cannot travel far or pass through skin, so poses no danger to humans unless ingested. If this happens, however, even a speck of polonium the size of the full point at the end of this sentence can carry over 3,000 times the lethal dose for humans. First discovered in 1898 by Marie Curie, who named it after her native Poland, polonium was isolated and used during the American Manhattan Project to develop an atomic weapon. Project scientists also tested polonium's effects on volunteers with incurable diseases; their autopsies revealed the results of polonium poisoning on the internal organs. The first confirmed case of polonium-210 being used to harm someone deliberately was in 2006, with Alexander Litvinenko (pages 156–59). Today, polonium-210 is difficult and time-consuming to extract, and only around 100 mg is manufactured in all of the world's nuclear reactors every year.

TOXIC EFFECTS:

Minuscule amounts of polonium can often be found in the body, but at higher doses there is virtually no more harmful substance in the world. If ingested or inhaled, even a millionth of a gram is usually enough to be fatal. Once it enters the bloodstream the radioactive particles attack the body at a cellular level and bombard the internal organs until one after another they stop working. Polonium then enters the bone marrow and causes the lymphatic system to shut down. Death is a natural conclusion to this highly charged attack. Because polonium is so toxic as well as being colourless, odourless and difficult to trace in the human body, it is sometimes known as the world's perfect poison.

SYMPTOMS:

Polonium is millions of times more toxic than hydrogen cyanide, and its effects are typical of radiation poisoning. These symptoms include nausea, vomiting, diarrhoea, hair loss and severe headaches, followed by liver and kidney failure. Death may take days or weeks to occur.

TREATMENT:

If a lethal dose of polonium has been recently ingested, it is thought possible to remove it by flushing out the stomach. However, once a lethal dose of the substance enters the bloodstream there is no stopping it. The alpha particles are absorbed by the body's cells and there they destroy all biological processes.

FAMOUS POISONINGS:

- Marie Curie won the Nobel Prize for discovering polonium but also died from her research, succumbing to leukaemia in 1934 caused by exposure to high-energy radiation.
- In 2004, Palestinian leader Yasser Arafat died suddenly from a mysterious illness that some believe was the result of polonium poisoning. A Swiss investigation did find traces of the poison on Arafat's personal belongings, but there has been no conclusive proof that it killed him.

OPPOSITE: Polish Nobel Prize winning scientist Marie Skłodowska Curie. Her discovery of radium and polonium directly contributed to her death.

Viktor Yushchenko Disfigurement

In 2004, Ukrainian presidential candidate Viktor Yushchenko returned home after a restaurant dinner and kissed his wife. She said his lips tasted metallic. Two days later, Yushchenko's body began swelling up.

As he was flown to an emergency clinic, Yushchenko's head grew in size and a terrible pain gripped him. Then, blackheads, cysts, pustules and lesions erupted over Yushchenko's face. This was chloracne, an unmistakable side effect of dioxin poisoning.

The dioxin attack on Viktor Yushchenko astonished world governments. This was a clear assassination attempt on a European politician running in a democratic presidential election. All eyes fell on his main opponent, Prime Minister Viktor Yanukovych, a pro-Russian, anti-European candidate who had the backing of the then president, Leonid Kuchma.

Kuchma's administration faced accusations of corruption and abuse of power as he pursued closer ties with Russia and Ukraine's economy floundered. Yushchenko had once been Kuchma's prime minister, but was ousted in 2001. This led to the formation of the Our Ukraine political coalition and a new style of liberal politics by Yushchenko, which was pro-Europe and encouraged dialogue among voters.

The campaign against Yanukovych was bitter and violent, but no one could have imagined that Yushchenko would be poisoned. A toxicologist's report found that the levels of dioxin in Yushchenko's blood were 50,000 times higher than in a normal adult. The quality of the dioxin was so pure that it could only have been made in a laboratory, the report concluded. Experts also noted that dioxin was a particularly suitable poison to use as it is tasteless and colourless, and its symptoms do not appear until days later. Few laboratories are equipped to detect dioxin in the blood.

For Yushchenko the crime was clear-cut: the three men with whom he had dinner on the night of the poisoning were working for the Russian government; they had tried to assassinate him, he said, by lacing his rice with dioxin. The Russian response, however, was that Yushchenko's condition had been caused by too much sushi and alcohol at the dinner.

Despite his scarred and pockmarked face, Yushchenko continued his presidential campaign. After a closely fought run-off in which Yanukovych was declared the winner, mass popular protests, known as the Orange Revolution, began. A second run-off called by the Supreme Court was then won by Yushchenko, who became president in 2005.

The investigation into the Viktor Yushchenko dioxin poisoning was still proceeding in 2019, with no deadline in sight. All requests to extradite the men who dined with Yushchenko on the night of the poisoning have been refused. When asked by a journalist if the poisoning went to the highest levels of the Russian government, Yushchenko replied: "My poisoning took place because I had started taking steps towards the European Union. We have a neighbour who does not want this to happen."

OPPOSITE: Viktor Yushchenko shows the disfiguring effects of dioxin in 2004. The condition, called chloracne, causes an eruption of cysts, acne, blackheads and pustules on the face, armpits and groin.

Dioxin

Dioxins are a group of chemical compounds usually created as by-products of herbicides and disinfectants. One specific dioxin has been synthesized for use as a poison. Its scientific name is 2,3,7,8-tetrachlorodibenzo-para-dioxin (2,3,7,8-TCDD), but it is commonly known simply as "dioxin".

OVERVIEW:

Dioxin is a dangerous carcinogenic, which caused skin lesions in German chemical industry workers exposed to the poison in the late nineteenth century. More recently, dioxin made up part of the herbicide and defoliant Agent Orange, used by the US military during the Vietnam War. More than 20 million gallons of Agent Orange were used to destroy tracts of jungle used as cover by the enemy, but led to serious illness in those who were exposed to the poison on the ground. It is estimated that over three million people suffered from deadly diseases caused by Agent Orange poisoning, including soft-tissue sarcoma, non-Hodgkin's lymphoma and Hodgkin's lymphoma. Birth defects in Vietnamese children are still occurring more than 43 years after the war's end. Dioxin accidentally released from a manufacturing plant in Seveso, Italy in 1976 contaminated the local population, resulting in infertility and a higher incidence of certain cancers in the years afterwards. Most examples of dioxin poisoning known to history have come from its exposure to skin and have been largely accidental. However, its use as a poison was highlighted by the attack on Ukrainian opposition leader Viktor Yushchenko in 2004 (see pages 162–63).

TOXIC EFFECTS:

Dioxin toxicity is notoriously difficult to measure: different amounts can have wildly differing results depending on a creature's metabolism. For this reason, it can take up to 5,000 times more dioxin to kill a hamster than a guinea pig. Some of the chemicals in dioxin bind with the DNA in a cell's nucleus, disrupting its ability to produce protein. This can lead to an organism's immunity being compromised and can cause certain types of cancer. In humans, the unifying feature of dioxin poisoning is that it produces the disfiguring chloracne that leaves the sufferer's face sallow, pockmarked and riddled with cysts.

SYMPTOMS:

Dioxin symptoms include irritation of the skin and mucous membranes following direct exposure to the poison. Internal symptoms can take several weeks to appear and can include chloracne, excessive hair growth, sensory impairments, muscle weakness and, at high doses, a gradual wasting away, followed by death.

TREATMENT:

There is no specific antidote for dioxin poisoning. Treatment of the symptoms includes stomach flushing and administering of activated charcoal for intestinal evacuation.

FAMOUS POISONINGS:

- In 1997, five people in Vienna were poisoned by dioxin while at work in a textile factory. One of the women recorded one of the highest levels of dioxin ever recorded in the human body: 144,000 picograms TCDD/g blood fat, or a calculated body burden of 1.6 mg TCDD. Miraculously, the women survived, but they suffered chloracne for many years afterwards.

OPPOSITE: A US plane sprays Agent Orange on Vietnamese jungle. The dioxin-carrying substance was used in the herbicidal warfare programme, Operation Ranch Hand.

Salisbury's Sergei and Yulia Skripal

When Sergei and Yulia Skripal were found slumped over on a park bench it was assumed they had taken a drug overdose. No one would believe that the sleepy town of Salisbury in Wiltshire had just become the centre of a geo-political thriller.

The Skripals had fallen victim to a covert assassination attempt by Russian operatives. They had been poisoned by a military-grade nerve agent that soon the world would know by name: novichok.

Eyewitnesses in Salisbury on 4 March 2018 describe Yulia Skripal as foaming at the mouth with her "eyes open but completely white". After the pair were airlifted to Salisbury District Hospital, it emerged that Sergei and his daughter Yulia had been poisoned by the extremely rare nerve agent novichok. More compellingly still, Sergei happened to be a former Russian intelligence officer and double agent for Britain's secret service.

Sergei Skripal had moved to England permanently after a spy swap with Russia in 2010. At the time, he was serving a 13-year sentence in a Russian penal colony for high treason. He had been caught selling state secrets to the UK's Secret Intelligence Service (MI6) and was believed to have blown the cover of over 300 Russian agents. After the Skripal poisoning, *The New York Times* reported that Sergei was retired but "still in the game".

According to local accounts, Sergei was a polite 66-year-old gentleman living out a rather ordinary retirement in Wiltshire. His 33-year-old daughter Yulia had flown in from Moscow the day before,

OPPOSITE TOP: CCTV images show the Skripals just minutes before the attack.

OPPOSITE: A policeman guards the Salisbury crime scene in the Skripal poisonings of 2018.

and they had gone for a drink and Sunday lunch in Salisbury town centre. They left the restaurant at 3.35 pm and were found unconscious on the park bench at 4.15 pm. The Skripals had been contaminated with such a high amount of novichok that it hospitalized the policeman who was first on the scene.

As the policeman and the Skripals fought for their lives, a police cordon was put around the Skripal home, and areas of central Salisbury were sealed off. With the influx of ambulances, police tents and investigators dressed in forensic suits, Wiltshire began to resemble a setting for a science-fiction disaster movie. When the poison and possible motive were revealed, the UK government accused Russia of attempted murder. Its accusation was supported by 28 allied countries, which expelled 153 Russian diplomats.

The evidence against Russia was indisputable: the novichok was identified as a strain called A-234, closely connected to the A-232 developed in Russia. Only a leading official would have the authority to approve the release of this novichok to be used in a cross-border assassination. So who, then, were the assassins?

Many assumed that an assassination attempt by Russian operatives would be a high-level affair using futuristic technology and the latest cloak-and-dagger techniques. It was a surprise, then, the two men responsible had simply flown in on a passenger airliner from Moscow a few days earlier and posed as tourists visiting Salisbury Cathedral. With them was a small bottle of perfume containing novichok.

Images obtained from CCTV cameras show the men arriving at London's Gatwick airport and then catching a train to Salisbury. The pair are seen walking towards the Skripal home, although they later claimed to have visited the cathedral. There are no CCTV cameras in the street where Sergei Skripal lived, but forensic evidence revealed that the novichok was sprayed on the outside handle of his front door. By the time the Skripals were found on the Wiltshire park bench, the assassins were in London, getting ready to board their plane to Moscow.

According to UK customs, the assassins had been travelling under the names Alexander Petrov and Ruslan Boshirov. A warrant for their arrest was issued in case they were found in Europe; however, there was a further twist when the men appeared on Russian state television to proclaim their innocence. They had been interested, they said, in Salisbury Cathedral's spire, one of the highest in Europe.

However, the interview only served to expose the men. One of them was identified by the investigative website Bellingcat as Colonel Anatoliy Chepiga, a veteran of the war with Chechnya and recipient of the order "Hero of the Russian Federation", the country's highest award. The other suspect was identified as Colonel Alexander Mishkin, a doctor in Russia's military intelligence service, the GRU.

While this went on, the Skripals had regained consciousness, stabilized and been released from hospital. Their present whereabouts has since been unknown, but in a television statement Yulia declined apparent help from the Russian government and asked that she and her father be left alone to convalesce. Russia accused the UK government of depriving the Skripals of their liberty and suggested that they be returned to Russia.

There was a final twist in the poisoning saga when a couple from Amesbury, seven miles north of Salisbury, used perfume from a discarded bottle. Fifteen minutes after applying the perfume, Charlie Rowley and his partner Dawn Sturgess fell ill. Sturgess had sprayed some of the perfume on her wrists, which Rowley later described to be an oily substance "that didn't smell of perfume". A friend found the pair at their home, Sturgess foaming at the mouth and Rowley rocking against the wall and "sweating, dribbling and making weird noises". After Sturgess later died in hospital, the autopsy showed novichok in her system. Despite the attempt to assassinate Sergei Skripal with the poison, Sturgess remains the only fatality from the attack.

RUSSIAN DENIALS

The assassination attempt by Russian agents on British soil led to the greatest diplomatic falling out between London and Moscow since the Cold War. Russia has consistently denied involvement, instead suggesting that Britain had fabricated the attacks to undermine Russian interests. In an interview, President Vladimir Putin said the assassins were Russian citizens who did not have criminal records. State television channel Channel One Russia joined in the denials by saying that the assassins had in fact been sports nutritionists visiting Salisbury to look for nutritional products. A monologue by one news anchor warned of the hazards of being "a traitor to the motherland" and noted that England had proven to be a particularly dangerous destination. "Maybe it's the climate, but in recent years there have been too many strange incidents with a grave outcome," she said. The examples of sudden, unexplained deaths of Russian nationals on UK soil include Kremlin whistleblower Alexander Perepilichny, Putin critic Boris Berezovsky and former spy Alexander Litvinenko (page 156–59).

The factor that unites this roll-call of victims, and others such as the Skripals, who survived, is their role as dissenters in Vladimir Putin's Russia.

ABOVE: *The box that contained novichok-contaminated perfume.*

OPPOSITE: *Colonel Anatoliy Chepiga and Colonel Alexander Mishkin defend themselves on Russian state television under the aliases Alexander Petrov and Ruslan Boshirov.*

Novichok

Novichok, meaning "newcomer" in Russian, is a term for a series of nerve agents developed from the phosphorus-oxygen-fluorine core of older nerve agents such as sarin and soman. It is similar in nature to the nerve agent VX.

OVERVIEW:

The Soviet Union first developed novichok in the 1970s under the secret military programme "Foliant". Its existence was revealed in the 1990s by Russian scientist Vil Mirzayanov after he defected to the US. Mirzayanov explained that batches of novichok had been developed in the secret Nukus chemical weapons facility in Uzbekistan. It was considered a valuable weapon by military leaders because it was unknown and therefore would escape detection by international inspectors, unlike older nerve agents such as VX. It was later confirmed that tests of the poison's efficacy were carried out on dogs.

Odourless and colourless, novichok is administered as a liquid, powder, gas or aerosol, as in the case of the Salisbury Poisonings (see pages 166–69). The 100-ml perfume bottle later discovered in Salisbury was calculated to contain enough novichok to kill thousands of people. Novichok in larger quantities could be released via shells, bombs or missiles.

In 2018, Germany's secret service acquired a sample of Soviet-developed novichok and shared it with its partners in Europe, Britain and the US. Novichok has subsequently been developed in some Western countries for test purposes. Although the effect of a city-wide novichok attack would be catastrophic, the toxin is no longer an anonymous weapon unknown to defence networks in developed countries.

TOXIC EFFECTS:

Like most nerve agents, novichok acts on the molecule acetylcholine, which enables messages to pass between nerves. Once it has done its job, acetylcholine is broken down by an enzyme called acetylcholinesterase. Without this breakdown, the muscles contract uncontrollably, or begin to spasm. This leads to respiratory failure and cardiac arrest. As a result the lungs fill with fluid, which eventually causes the victim to drown if they do not die from a heart attack first.

SYMPTOMS:

The first symptoms of novichok poisoning are severely constricted pupils, interrupted breathing, high heart rate and profuse sweating, followed by nausea, vomiting, diarrhoea, convulsions, loss of consciousness and death.

TREATMENT:

There is no antidote, but the drug atropine can help to stop the action of the poison by blocking the acetylcholinesterase enzyme. However, atropine can only be effective if the victim receives it soon after being poisoned with novichok. By further relieving respiratory muscle paralysis, thus enabling the victim to breathe, and removing liquid from their lungs, it is possible to save someone poisoned with novichok.

FAMOUS POISONINGS:

- In 1995, an early variant of novichok was used to poison Russian banker Ivan Kivelidi and his secretary Zara Ismailova. Kivelidi's former business partner Vladimir Khutsishvili was later convicted of the murders. It was thought that he illegally obtained the poison from an employee of the State Research Institute of Organic Chemistry and Technology, which was involved in the development of novichok.

OPPOSITE: The perfume bottle used in the Salisbury novichok attacks.

Index

(Page numbers in *italic* refer to photographs, illustrations and captions)

A

acetylcholine 31
aconite (*Aconitum* species) 10, *36*, *37*, 37
acqua tofana 55–6
actium 22–3
Affair of the Poisons 75–7
Agent Orange 15, *164*, 165
Agrippina 33–4, *34*, 35
alchemy 41, 46, 59, 75
Ancient Egypt 8–9, 11, 12, 16, *24*
anthrax (*Bacillus anthracis*) 13, 148, *148*, *149*, 150, *150*, 152, 153, *153*
antimony 12, 103
Antony, Mark 22, 23, 85
aphrodisiacs 11, 70
Arafat, Yasser 161
arrow-tipped poisoning 16, *16*, *17*, 37
arsenic 8, 12, *12*, 41, 53, *86*, 88, *89*, 90
atropine 31, 37
attack poisons 11
Aum Shinrikyo *134*, 135–6, *135*, 136
Auschwitz-Birkenau 13
Australian thallium craze 147

B

Babington, Sir Anthony 63
Bacon's Rebellion (1676) 82, *82*, 83
Banquet of Chestnuts 50
bear's foot *see* aconite
belladonna (*Atropa belladonna*) *10*, 16, *30*, 31
Black Death 13
black magic 46, 66, *66*, *67*, 75, 77
Blandy, Mary 86, *87*
blister beetle (Meloidae family) 11, *72*, 73, *73*
Blowing Up Russia (Litvinenko) *158*, 159
Borgia poisonings 48, 49–50, *49*
Bradford Sweets Poisoning 92, *92*, 93
Brinvilliers, Madame de *74*, 75, *75*
Britannicus 31, 33–4
bubonic plague 13, 81

C

cane toad (*Rhinella marina*) 11
castor bean (*Ricinus communis*) *132*, 133, *133*
central nervous system 21, 31, 39, 43, 65, 70
Chambre Ardente 75, 77
charcoal treatment 37, 65, 85
Cheema, Lakhvinder ("Lucky") 37
Chepiga, Col. Anatoliy 168, *168*, 169
Chinese medicine 16, 37
chlorine *112*
chloroform 107, *110*, 111, *111*
Christie, Agatha 6
chronic poisoning, defined 15
Cleopatra 8–9, 11, 22–3, *24*–5
cobras 11, *22*, 23
Coca-Cola *142*, 142
Cold War 13, 112, 130, 169
Collier, John 50
coniine 21
Cotton, Mary Ann 94, *94*, 95
Cream, Dr Thomas Neill *106*, 107–9
Crippen, Cora ("Belle Elmore") (née Turner) *126*, 127–9, *128*
Crippen, Dr Hawley *126*, 127–9, *129*
Curie, Marie Sklodowska *160*, 161, *161*
cyanide 6, 9, 13, 88, 112, 117, *119*, 120, *122*, 124, 125, 154

D

Darwin, Charles 53
Daschle, Tom 150
Datura stramonium 84, 85, *85*
David, Jacques-Louis *18*, 19
deadly nightshade *see* belladonna
death cap (*Amanita phalloides*) 11
death cherries *see* belladonna
The Death of Cleopatra 24–5
The Death of Socrates *18*, 19
devil's herb *see* belladonna
devil's porridge *see* hemlock
devil's trumpet *see Datura stramonium*
diazepam 21
Dickens, Charles 65, 103
Dio, Cassius 23, 34
dioxin 15, 162, *164*, 165, *165*
Dirlewanger, Oskar 105
doe-eyed, defined 31
doomsday cults 135–6
Dr Mackenzie's Arsenical Soap *12*
dwale *see* belladonna

E

Ebers Papyrus 8–9
Egyptian cobra (*Naja haje*) 22, 23
elapids 14
Elizabeth I 62
emerald-green blister beetle (*Lytta vesicatoria*) 72, 73, *73*
ergot (*Claviceps* genus) 11, 80, 81, *81*
exotoxins 13
exposure, routes of 14
extermination camps 13, *13*, 112, *116*, *117*, 118

F

female healers *10*
Flavor Aid 117, 123, 124, 125
fly agaric (*Amanita muscaria*) 11, *11*, 33
flying sensations *10*, *11*, 66
Fowler's Solution 53, 90, *90*
François, Donatien Alphonse *see* Sade, Marquis de
French school of poisoners 61
fungi 11, 13

G

Gaza War (2008–09) 99
Glass of Wine with Caesar Borgia 50, *51*
Golden Age 8, 88–111
Goreck, Susan 142, *142*, *143*
Great Wall of China *see* Qin Shi Huang

H

Hardaker, William ("Humbug Billy") 92

Heimskringla 81
hellebore 16
hemlock (*Conium maculatum*) 10, 20, 21, *21*
hemorrhagic gastritis 86
henbane (*Hyoscyamus niger*) *10*
Himmler, Heinrich 118, *119*
Hindu medicine 37
Hitler, Adolf 118, *118*
Holocaust 118
Homer 16, *16*, *17*
homophobia 120
hydrogen cyanide *see* cyanide
hyoscine hydrobromide 127, 128, *129*
hyoscyamine 31

I

Iceman *see* Kuklinski, Richard
Iliad (Homer) 16
immunity 8, 28, *29*
inheritance powders 66, 77
insect poisons 13
insoluble poisons, defined 14
intoxicants 11, *11*
Islamic scholarship 28

J

Jack the Ripper 107, 108, *108*, 109, *109*
Jamestown Poisonings 82
Jamestown weed (*see also Datura stramonium*) 82, 85
Japanese teenage poisoner 144
Jewish people, and Nazi Germany 13, 105, 112
jimsonweed *see Datura stramonium*

Jones, Jim 112, 123–5, *123*
Jonestown Massacre 117, 123–5, *124*, *125*

K

Kuklinski, Richard 154, *154*, *155*

L

La Bourboule *88*
La Cantarella 49
La Neve *see* Neave, Ethel
La Voisin 9, 66, 75–7, *76*
Labour movement 96, *97*
Lafarge poisoning 53
Lambeth Poisoner *see* Cream, Dr Thomas Neill
laudanum 65
laxatives 12
LD50 15
lead and lead poisoning 12, 38, 39, *39*
Leahy, Patrick *150*
lethal dose, defined 15
lethal injections 124
Litvinenko, Alexander 156, 157–8, *157*, *158*, *159*
Locusta 31, *32*, 33–4, *33*
London Pharmacopoeia 28
Lopez, Roderigo 62, *62*
Louis XIV 9, 66, 69, 75–7, *77*
Love Potion (1903) 57
LSD 136

M

Mad Hatter's Tea Party *42*
magnificences 46
Majdanek *116*, *117*
man-made poisons 13

Markov, Georgi 130, *130*, 131
Marsh, James 88, *89*
mass suicide 26–8, 112, 117, *124*, *125*
matchstick girls 96, *97*
Maybrick, Florence Elizabeth 90, *91*
Medici, Catherine de' 46, 49, *58*, 59–61, *59*
medicine, as poison 8
medieval period 46–65
mercury (Hg) and mercury poisoning 12, 41, 43
metals *see* lead and lead poisoning; mercury
Metamorphoses (Ovid) 16
Middle Ages 31, 37, 46–65, 81
Mishkin, Col. Alexander 168, *168*, *169*
Mithridates 8, 26–8, *26*, *27*, *29*
Mithridatium (Mithridate antidote) 8, 26, 28, *28*
monkshood *see* aconite
Montespan, Madame de 75–7, *77*
Monvoisin, Catherine *see* La Voisin
Morgan, Evelyn de *57*
moulds 11, *80*, 81
Moumié, Félix-Roland 147
Mozart, Wolfgang Amedeus *54–5*, 55–6
mushrooms *see* fungi
muskrat weed *see* hemlock
mustard gas 112, *112*
My Fifteen Lost Years (Maybrick) *91*
mycotoxins 11

N

Neave, Ethel ("Le Neve") *126*, 127–9, *129*
Nero 28, 31, 33–4, *34*
neurotoxins 21
nicotinic acetylcholine receptors 21
9/11 148, 150
novichok 8, 13, 167–8, *169*, *170*, 171, *171*
nux vomica tree (*S. nux-vomica*) *104*, 105, *105*

O

Odyssey (Homer) 16, *16*, *17*
opium 16, 21, 65
opium poppy (*Papaver somniferum*) *64*, 65
Orfila, Mathieu 88
Ovid 16
oxygen, as poison 8

P

Palestinians, Israeli military poisoning of 99
Palmer, William *100*, 101–3, *101*, *102*
Paracelsus 8, 12, *47*
paralysis 8, 11, 21, 23, 43
Phaedo (Plato) 19
pharmaceutical poisoning 12, *12*
pharmacologically active, defined 10
Phipps poisonings 43
phosgene 112
phosphate rock *98*, 99, *99*
phosphorus (*see also* white phosphorus) 9, 49, 99, 171
phossy jaw 96, *96*, 99
Physostigmine 31
phytotoxins 10
plant poisons 10, 14
Plato 19
Pliny the Elder 23, 26, 28
Plutarch 22, 23
poison:
    in fiction 6–7, 16, 130
    science of 14
poison hemlock (*Conium maculatum*) 10, *20*, 21, *21*
poison parsley *see* hemlock
poisonous alkaloids 37
polonium 13, 157–8, *157*, 161
Pompey, Gnaeus 26, 28
potassium chloride 124
potassium cyanide *see* cyanide
Prince of Poisoners *see* Palmer, William
prostitution 50, 59, 66, 69–70, 73
protein synthesis 13
Ptolemy 26
Punch 92
Putin, Vladimir 157, 158, *158*, 169

Q

Qin Shi Huang 12, *40*, 41, *41*
Queen of Poisons *see* aconite

R

Rasputin, Grigori ("Mad Monk") 114–15, *114*, *115*
rat poison *9*, 13
reindeer *11*
Renaissance 28, 31, 46–65
revolutionary suicide 123, 125
ricin 133
Rixens, Jean André *24*
Rodine *9*
Roman Empire 9, 16, 22–3, 26–8, *29*, 33
Roman History (Appian) 28
rye ergot (*Claviceps purpurea*) *see* ergot

S

Saddam Hussein 139, 147
Sade, Marquis de 11, 66, *68*, 69–71, *69*, *70*
Saint Bartholomew's Day Massacre (1572) *61*
St Vitus's Dance 81
Salem Witch Trials 11, 78, *78*, *79*
Salisbury poisonings 8, *166*,

167–9, *167*
Santa Claus 11
sarin gas 13, 135–6, *138*, 139, *139*
scorpions 11
Secretissima (The Ten) 60
serial killers 55–6, 65, *94*, 108, 111, 140, *140*, 144, *144*
Sforza, Count Francesco, Duke of Milan 46, 49
Shakespeare, William 23, 31, 37
shamanism 26
Shipman, Harold Frederick 140, *141*
Shoko Asahara 135–6, *135*
Singh, Lakhvir 37
Skripal, Sergei 8, 167–8, 169
Skripal, Yulia 8, 167–8, 169
snakes and snake bites 11, *14*, 22, *22*, 23
soap 12
Socrates 16, *18*, 19, 21
soluble poisons, defined 14
Solzhenitsyn, Aleksandr 133
Spanish fly *see* emerald-green blister beetle
spiders 11
Stashynsky, Bohdan 117
stink weed *see Datura stramonium*
Stride, Elizabeth *108*
strychnine 6, 105
suicide 11, 21, 26–8, 112, 117, 123, *124*, *125*
sulphur mustard 112
Sylvestre, Joseph-Noel *33*
Syrian Civil War 139, *139*

T

Tacitus 34, 73
taipans 11
Talbot Arms Inn *102*
Teacup Poisoner *see* Young, Graham
10 plagues of Egypt 153

Terracotta Army *41*
thallium (Tl) 12, 142, *142*, 144, *144*, *146*, 147, *147*
Thirty Tyrants 19
thorn apple *see Datura stramonium*
"three kingdoms of nature" 46
toadstools *see* fungi
Tofana, Giulia 9, 55–6, *57*
Tofana Trap 55–6
toilet soap *12*
Tokyo subway poisonings 13, 135–7, *136*, *137*
Top Top Secret (The Ten) 60
toxic alkaloids 21, 31
toxikón/toxikós 16
toxin, defined 16
Trepal, George 142, *142*
tripping 11
Triumph poisonings 43
Trojan War 16, *16*, *17*
Turing, Alan 120, *120*, *121*
2,3,7,8-tetrachlorodibenzo-paradioxin (2,3,7,8-TCDD) *see* dioxin
Tylenol killings 117

U

umbrella death 130, *130*

V

Venetian Council of Ten ("The Ten") 46, 60, *60*
Venetian treacle 28
venom 11, *14*, 22, 23
Vienna dioxin poisonings 165
Vietnam War 15
vinegar 16
vipers 11, *14*, 28
VX 171

W

Washinton anthrax attacks *146*, *147*, 150, *150*, *151*

water, as poison 8
weedkillers 13
Wetterhahn, Karen 43
white arsenic (*arsenic trioxide*) (*see also* arsenic) 53
white phosphorus (*see also* phosphorus) 96, 99
Winnett, Tony 43
witch mythology *10*, 46, 66, *67*
wolfsbane *see* aconite
Woodruff, Angela 140
World War I 112, *112*, *113*
World War II 99, 112, 117, 118, 120, 129
World War III 136
Wu Zetian 44, *44*, *45*

Y

Ying Zheng *see* Qin Shi Huang
Young, Graham 144, *144*, *145*
Yue Fei *44*, *45*
Yushchenko, Viktor 148, 162, *162*, *163*

Z

Zimbabwe animal poisonings 117
zinc *12*
Zopyrus 26
Zyklon B 13, *13*, 112, 117

Credits

The publishers would like to thank the following sources for their kind permission to reproduce the pictures in this book.

AKG-Images: DeAgostini/A.Dagli Orti 73

Alamy: AF Fotografie 68; /Archive PL 96; /Artokoloro Quint Lox 28; /Stefano Bianchetti 74; /Chronicle 10, 17, 44, 87, 109; /Dorset Media Service 170; /Geoz 98; /Florilegius 104; /John Frost Newspapers 118; /F.Martinez Clavel 72; /History and Art Collection 57; /ImageBROKER 14; /Lanmas 67; /Lebrecht Music & Arts 62; /Newscom 134, 169; /Pictorial Press 97; /UtCon Collection 71; /Bruce Yuanyue Bi 130

Bridgeman Images: 11, 20, 27, 102; /Archives Charmet 32; /Granger 63; /Look and Learn 29, 51; /Purix Verlag Volker Christen 31

Getty Images: 131, 141; /Araldo de Luca/Corbis 35; /Bentley Archive/Popperfoto 126; /Bettmann 119, 154; / Bride Lane Library/Popperfoto 123; /BSIP/UIG 22; / The Cartoon Collector/Print Collector 93; /DeAgostini 60; /Express Newspapers 108; /FBI 151; /Fine Art Images 45, 47, 48; /Hoang Dinh Nam/AFP 15; /Efired 41; /Florilegius/SSPL 64; /Granger 77; /Acey Harper/The LIFE Images Collection 142, 143; /Julien M. Hekimian 70; /Heritage Images 40; /Nobori Hashimoto 137; /Hulton Archive 63, 128; /Illustrated London News/Hulton Archive 103; /Imagno 24-25; /Wojtek Laski 115; /Markus Leodolter/AFP 163; /Hasan Mohamed/AFP 138; /Matthew Naythons/The LIFE Images Collection 122, 123; /NY Daily News Archive 155; /New York Times Co./Neal Boenzi 124; /Ira Nowinski/Corbis/VCG 116; /Phas/UIG 77; /The Print Collector 91; /Popperfoto 129; / Prisma/UIG 36; /Prisma Bildagentur/UIG 61; /Smith Collection/Gado 152; /Dick Swanson//The LIFE Images Collection 164; /TASS 168; /Ullstein Bild 114; /Alex Wong 149; /SSPL 38, 65, 110, 120; /Universal History Archive 12, 54-55, 64, 88, 160; /Natasja Weitsz 156

Greenwich Heritage Centre: 89

Mary Evans Picture Library: Photo Researchers 86

Metropolitan Museum Of Art, Catherine Lorillard Wolfe Collection, Wolfe Fund, 1931: 18

Private Collection: 69, 90, 95

Science Photo Library: 146; /Carlos Clarivan 157; /Natural History Museum, London 52; /Dr. Mark J. Winter 144

Shutterstock: 113, 166; /Amoret Tanner Collection 9; /Design Pics Inc 58; /Eye of Science 80; /Alistair Fuller/AP 159; /Granger 79, 121; /MoreVector 13t; /Morphart Creation 42; /Stanislav Samoylik 13b; / Geoffrey White/ANL 145

Topfoto: 106

Wellcome Library: 100

Every effort has been made to acknowledge correctly and contact the source and/or copyright holder of each picture and Carlton Books Limited apologises for any unintentional errors or omissions, which will be corrected in future editions of this book.